Gesunde Ernährung

Schriftenreihe der Dr. Rainer Wild-Stiftung

Healthy Nutrition

Book series edited by the Dr. Rainer Wild-Stiftung

Springer
Berlin
Heidelberg
New York
Hong Kong
London
Milan
Paris
Tokyo

P. Belton · T. Belton (Eds.)

Food, Science and Society

Exploring the Gap Between Expert Advice and Individual Behaviour

Springer

Prof. PETER S. BELTON

School of Chemical Sciences
University of East Anglia
Norwich NR4 7TJ
United Kingdom

Dr. TERESA BELTON

School of Education and Professional Development
University of East Anglia
Norwich NR4 7TJ
United Kingdom

ISBN 3-540-43743-6 Springer-Verlag Berlin Heidelberg New York

Library of Congress Cataloging-in-Publication Data
Food, science and society : exploring the gap between expert advice and individual behaviour / P. Belton ; T. Belton (eds.)
p. cm. – (Gesunde Ernährung)
Includes bibliographical references and index.
ISBN 3-540-43743-6 (alk. paper)
1. Food adulteration and inspection. 2. Food-Social aspects. 3. Risk assessment. I. Belton, P. S. II. Belton, T. (Teresa), 1952- III. Series.

TX531 .F5687 2003
363.19'26–dc21

Springer-Verlag Berlin Heidelberg New York
a part of Springer Science+Business Media

http://www.springer.de

Printed in Germany

Production Editor: Renate Albers, Berlin
Typesetting: Data conversion by Fotosatz-Service Köhler GmbH, Würzburg
Coverdesign: Struve & Partner, Heidelberg

Printed on acid-free paper SPIN: 10992371 52/3111 - 5 4 3 2 1

Preface

There is widespread concern amongst consumers about the safety and acceptability of food. At the same time industry, regulators and scientists find themselves frustrated that their advice is ignored or dismissed for reasons that they do not understand or that appear to them to lack credibility. There is clearly a communication gap between consumers and many food professionals. There is no easy way to bridge this gap. However, we believe that the situation can be improved by developing a more empathetic and open approach to the issues. In order to improve dialogue scientists must re-examine their position and the limitations to a purely technical approach to food issues.

Science is no longer accorded the authority it once enjoyed. It is one among a number of contending stances. It cannot expect its claims to objectivity and value free knowledge to be universally accepted. At the same time science does have much of relevance to say about the scientific and technological issues facing society today, and its contribution to these debates should not be marginalised. The problem for the scientist is how to engage in debate in a way in which the contribution of science can be weighed together with the, sometimes unspoken, contribution from other perspectives as well as social and cultural factors.

The book falls into three sections. The first section examines the general issues of the roles of science, culture and risk perception in relation to food at the beginning of the 21st century. The second contains two chapters which examine two particular areas. One is the important role of the mass media in questions of food risk and policy. The other is a detailed analysis of attitudes to eating fruit and vegetables, used as an exemplar of the need to understand such matters in a broad context.

The third section describes three practical experiences at the interface between scientists and lay people, in policy-making, agricultural practice in a traditional culture and awareness raising. These three chapters offer accounts of the two-way nature of the communication process and of how attempts may be made to bridge the gaps in a variety of social and cultural environments.

Norwich, June 2002 — Peter S. Belton and Teresa Belton

Foreword

In each individual's life, eating and drinking play a central role. All over the world, people consume food and drink, usually on a daily basis. Practiced from the beginning, every day contains any number of mostly spontaneous and arbitrary encounters with food and drink. These encounters are not solely of a physiological nature but are also always marked by norms and values of society and by personal likes and dislikes. Hence, regarding their own food, every individual is an expert from very early on. It is usual to talk during meals as well as about meals, for example, what was eaten, when and where, how it was prepared, and what it was like.

Generally, science choses a very different approach to eating. Instead of subjective experiences, "objective" examinations of certain issues, and observations and calculations are made, compared and discussed in context with other scientific data. Scientific expert knowledge develops only after a long period of purposeful dealing with the relevant applicable methods. The scientific discourse about food poses questions such as how is food eaten, which food and ingredients are consumed, how nutritional is it, and on what basis are choices made. Based on scientific facts, the individual is often instructed as to how and what they should eat and why. It is not surprising that mutual understanding is hard to come by – here, different worlds meet.

The 8th volume of the series by the Dr. Rainer Wild-Stiftung is concerned with this problem. Thus, it fits in well with the work and goals of the Heidelberg foundation for a healthy nutrition, which strives to contribute to the holistic understanding of nutrition. To this end, the foundation addresses scientists and mediators in the food sector and encourages them to look beyond the obvious. Details regarding the philosophy and activities of the Dr. Rainer Wild-Stiftung can be found in the back of the book.

The topic of the book directly picks up the thread from previous volumes of the series that dealt with consumer uncertainty (K. Bergmann, *Dealing with Consumer Uncertainty: Public Relations in the Food Sector*, 2002) and with the public debate about genetically modified food (L.H. Grimme and S. Dumontet, eds., *Food Quality, Nutrition and Health*, 2000). It meets Peter Belton's demand to think about the position of science in a post-modern society (P. Belton, 'Nutritional Science in Global Perspective', in G.U. Schönberger and

U. Spiekermann, eds., *Die Zukunft der Ernährungswissenschaft*; 2000, published in German only).

We would like to thank the editors, Teresa and Peter Belton, for allowing us to publish their book as part of this series. We are pleased that communication regarding this publication was not only considered theoretically, but was also carried out practically and smoothly. Furthermore, we would like to thank all authors for embracing the topic and for pursuing it in their contributions. Last but not least, a thank you to the team of the Springer-Verlag for their faithful co-operation and assistance at all times.

Dr. Rainer Wild-Stiftung | Heidelberg, May 2002

Contents

Science in the Post Modern World

1

PETER S. BELTON

1.1 Introduction

Twenty-first century scientists and technologists find themselves in a curious position. Whilst they can justifiably claim that the scientific method has led to huge advances in the quality of life for at least part of the world's population, they are at the same time assailed from a variety of directions. Science, which term I shall use as shorthand for science and technology, and scientific values, are blamed for many of the ills of the world (Appleyard 1993) and scientists are held in such a position of increasing distrust that Durant has been moved to comment (Durant 1997) "In recent years it has become common for technical risk assessments to be received, not with public indifference or incomprehension, but rather with public incredulity".

Science is becoming increasingly unpopular as a subject for study at school and university; even the fundamental intellectual claims of the scientific method, as objective and related to some external reality, are being radically challenged. In this context it is not surprising that food science and the value of the contribution of food scientists to the debates on food safety and acceptability are under question.

In this chapter I shall consider the nature of food science in the context of the nature of food, and the extent to which science can be used as an objective tool for the formation of policy. In doing so it will be necessary to discuss the way in which scientific information can be shared with the members of society who are not themselves specialists in the scientific disciplines, and the limits to science itself.

1.2 Food Science

There is a sense in which the phrase "food science" is a contradiction in terms. Food is a cultural and social concept, what constitutes food is determined by the social and cultural milieu of the potential consumer. We learn about food at the breast. Suckling contains elements of both the sexual and the satisfaction of hunger; food and oral contact are often associated with sexual behaviour in adulthood. Food has a role in social and familial bonding: typically we offer food as a symbol of friendship, and in modern Britain dinner parties and lunch clubs are a common means of social exchange. In a family, eating together is one of the ways in which relationships are maintained. We also offer food as a sign, as a gesture of hospitality. Food also may be used as means of defining social status and prestige; the quality and cost of food offered and the way in which it is presented can be used to send messages about relative social status and the wealth and good taste of the host.

In almost all religions food has a role in rites and is subject to a variety of proscriptions. In some Christian churches the sacrament is at the core of religious ritual, within it the body and blood of Christ, in the form of a wafer and wine, are ritually consumed. The meanings of this process are far from any material assessment of the nature of the food consumed. In this ritual the sacramental value of the food is paramount.

Food taboos abound. These may be religious, for example, the banning of the consumption of pork, or they may be social and cultural: in the UK the consumption of horse and dog meat is frowned upon – dog meat is not generally regarded as appropriate food, whilst in other cultures they are regarded as tasty and nourishing. Many vegetarians are people who find the consumption of meat repellent. There is no "right" or "wrong" list of foods, even proven risks of narcotic and/or toxic effects in, for example, alcoholic drinks and the Japanese fugu fish, are not a bar to consumption in some cultures.

A working definition of food might therefore be: *Food is organic matter, usually slightly decayed, but sometimes live, that someone wants to eat.* The point about this definition is that there are two components to it: matter and human motivation, the proper study of the former being physical and biological science, and of the latter being psychology, sociology and cultural studies.

The idea was summed up eloquently and succinctly by Terry Eagleton (Eagleton 1998), who wrote: *"Food is an interaction not an object."*

The reduction of food merely to object is therefore erroneous; it is possible to study only one aspect of food, its material properties, within the remit of physical and biological science. Any attempt to persuade consumers of food that some particular course of behaviour is desirable, based merely upon the one dimensional view derived from traditional food science, is unlikely to succeed unless it happens to accord with the other factors that play a critical role in food consumption. The need to take account of non-scientific factors in science communication is explored further in the next sections.

1.3 Communicating Science

We live in a society in which much of the change is driven by scientific advance and in which science is seen by many as a means by which value-free, objective knowledge may be obtained. However science is not the only driver of change nor is it the only form of knowledge. Scientists do not generally claim special knowledge in moral and social matters such as divorce law. Nevertheless there is a view among some scientists and technologists that when the subject of debate does include a scientific element there is some sense in which the scientific view is the pre-eminent one and, if only the public understood the scientific facts the matter could be resolved. Such a view has been characterised as the "deficit model" (Irwin et al. 1996) or the "public ignorance" model (Irwin and Wynne 1996 a). Such a model embodies the notion that science itself is in some way value-free and outside the social structure to which the consumer belongs. Implicit in this it assumes that there is some separate class of individuals called consumers or the public that do not enjoy the comprehensive and sufficient world view enjoyed by a class of individuals known as scientists. Following Irwin and Wynne, (1996 b), the deficit model may be characterised by a set of distinct but not self-evident assumptions. Some of these are:

1. The public addressed by the purveyor of scientific knowledge is essentially one that does not have its own legitimate values and that the experts' values and beliefs are either superior to those of the public, or will be automatically accepted by them.
2. Ignorance of scientific information is a result of an intellectual vacuum or an inability to understand. It is essentially passive and does not result from any active response to a social position or relationship to scientific or technical institutions.
3. The lay public expects certainty and risk free environments. The apparent refusal to accept scientists' assessments of relative risk arises because there is no real understanding by the public that risk and uncertainty are intrinsic to every day life. Even when the reality of risk is accepted there is no understanding of the relative quantification of various risks leading to an inappropriate hierarchy of risk acceptability.

These assumptions are ones that are very comfortable for the scientist since they all imply that any problems arising with the communication of science and the failure of the public to accept and act on scientific advice arises because of defects on the part of the public. The scientist remains blameless as a neutral purveyor of truths.

In their book, *Fearing Food*, Morris and Bate (1999) set out to answer concerns about the effects of pesticides on food and the environment. In so doing they and their contributors offer clear examples of the use of the deficit model. For example Morris (Morris 1999a) quotes Julian Huxley in an introduction to a British edition of Rachel Carson's book *Silent Spring* as follows:

The present campaign for mass chemical control, besides being fostered by the profit motive, is another example of our exaggeratedly technological and quantitative approach.

Morris then comments:

I shall refrain from commenting on the pejorative description of the profit motive and the utterly spurious militaristic implication of talking about a campaign for mass chemical control, because they are merely incidental.

So, even when Huxley is clearly including arguments that are not technological and relate to views about the nature of the economic structure of society and the whole style of the deployment of the chemicals in the environment his views are explicitly dismissed as pejorative or spurious. The core assumption in this passage by Morris is that the only criterion for the assessment of any argument is the technological. This is a very clear example of the attitudes summarised above. Anybody, even a distinguished scientist and thinker, such as Huxley, who does not conform to the rule that science is the sole arbiter of such issues, is dismissed as including irrelevant material in their discussion. By this means their argument is focussed onto a narrow ground in which the experts will always be in the strongest position, they will have access to the resources, the literature and the specialist language that will enable them to prevail in any confrontation. The experts may then win the argument in their terms, but as good management practice has shown, winning an argument by denying others the ability to express their position fully does not amount to winning support for a point of view or enthusiastic cooperation. It is more likely to lead to discontent and surreptitious sabotage. A good decision and one that is likely to gain support is one that is felt to be fair by most of those involved.

Of course an exchange that moves away from the narrow scientific arena and begins to involve concerns about the economic system and matters of the organisation of society is a much more difficult one and one in which it is much harder to gain consensus, since within such a debate there can be no appeal to the apparently value-free arbitration of science. The temptation is therefore always to restrict the debate and the terms of reference to purely scientific matters. This also has the political advantage that it shifts the burden of responsibility from the policy makers and politicians to scientific experts. If things go wrong in a situation where there is a technical input to decision making, such as in the BSE crisis, politicians can claim to have been led by the scientists and distance themselves from the political consequences.

Apart from the difficult political implications of fully recognising the non-scientific elements in decision making there are practical considerations: regulators do need to make defensible decisions in realistic time scales, not everything can be the subject of lengthy debate about intangible, albeit important, factors. There is thus always going to be tension between the need to take action and the need to recognise and deal with all the factors that may be in play in the debate. This is essentially a political decision. It is not one that can be made on a scientific basis. If the only input into a debate involving technical issues is to be a scientific one it must be recognised that a political decision has been made to limit the debate. It is not possible afterwards to claim that the only factors that needed consideration were scientific ones, although it may be possible in some uncontroversial circumstances to claim that scientific factors were the only important ones.

Elsewhere Morris (Morris 1999b) writes:

Environmental and consumer organisations have for a long time attacked the use of pesticides, fertilisers and other aspects of intensive farming on the grounds that these technologies are bad for the environment and bad for our health. However their arguments have typically been informed by poor or partial understanding of the scientific evidence regarding the impact of these technologies.

This is an interesting passage since it assumes that judgements about the quality of things such as environment and health are to be made purely by recourse to science. According to this interpretation no human, social or cultural elements are involved. This would seem to run counter to much research which suggests that what might be good or bad for health, or indeed whether such things are affected by physical agency, are very much dependant on the social and cultural milieu (Caplan 1997). Similarly the concept of environment is an extremely complex one with many attributes (Soulé and Lease 1995). Even without the academic support, a moment's reflection about one's own feelings about one's own health and the environment would suggest that these are not viewed entirely objectively and that personal preferences play a major role.

The second sentence is also interesting; it seems to recognise that there has been a serious attempt to critically evaluate the scientific evidence but that this attempt has been inadequate. The attitude is exemplified by the chapter called "Pollution, Pesticides and Cancer Misconceptions" (Ames and Gold 1999). The chapter lists a range of misconceptions which are simply stated at the heading of each section, for example: "Misconception No 1: cancer rates are soaring". In most sections very few or no citations are given to works from which these misconceptions are derived and where such citations are given they are dismissed in a couple of sentences. The authors of this chapter are distinguished scientists and they may well have poor opinions of the science behind these ideas but the tenor is not that of a scientific debate, with its careful citations and its circumspect language. It is that of polemic. It dismisses the arguments put forward without the usual courtesies and thus implicitly reduces the respect in which the proponents of these ideas are held. This dismissal may express the feelings of those who are contemptuous of the arguments so demolished. However it is not likely to convince their proponents that they are being heard, or that their views are being properly considered, because they are not being shown to be inadequate by the normal standards of scientific argument. In effect they are not being taken seriously. The point here is not who is "right" or "wrong" but what is the nature of the communication that has taken place. Is it between the apparent protagonists or is it concerned with persuading some third party of the soundness of the participants' views? Is it then thus more akin to advocacy than then scientific debate?

1.4 Science as Advocacy, or not

Polemical approaches are not the sole prerogative of the scientific establishment. The Greenpeace website (Greenpeace 2001) has a document entitled "Great Myths of the Incineration Industry" which has a very similar style to the chapter by Ames and Gold. There is thus reciprocity between the two sides of the argument. In this context Yearley (1996) has commented on the way in which campaigning groups in the environmental field have used science. He points out that there is a fundamental ambiguity in their position in that many of the leading environmental problems can be seen as the outcome of science and technology but that in secular society appeal to scientific expertise is the principal form of legitimisation. Therefore there has been an increasing tendency to make use of scientific authority and expertise by all sides in debate involving technical issues.

We now seem, therefore, to be in a situation that whatever the actual political, social and cultural drivers might be, the argument must be fought in the discourse of science. This has meant that for both sides it has become necessary to find expert opinion or commission research which will support of their views. Given the dialectical nature of science this is not hard to do. However the clash is between parties that have much at stake. On the establishment side industries are concerned about return on investment and the maintenance of business, while governments are concerned about economic growth and the maintenance of political credibility. On the protesters' side there are concerns about the sustainability of the environment, the aesthetic and moral issues involved and the fact that campaigning bodies owe their continued existence to the belief of their supporters that they are effective and are responding to real concerns. The object of the debate is not to clarify the best state of scientific knowledge but to persuade third parties, be they the protagonist's own constituency, the public, politicians or the chairman of a public enquiry, of some point of view. The audience then begins to determine the parameters of the debate, and science has thus been moved from what it might claim as its natural position of neutral objectivity to one of advocacy. It has thus become a weapon with which to win a debate that has its roots elsewhere. Given this subtext it is not surprising that a common search for an objective truth, the supposed outcome of the scientific approach, is lost and accusations and counter accusations of bias and misrepresentation are frequent. Thus by becoming the tool by which claims and counterclaims are legitimised it has in some sense lost its legitimacy and become a political weapon. This has lead to the paradoxical situation in which the supposed arbiter of truth is no longer regarded as such by large parts of the public.

An alternative to the use of the scientific discourse is what Michael (1996) has called "discourses of ignorance". Here he claims that individuals may deliberately claim or ensure their ignorance of all or some science as a positive means of identifying their situation and their attitude to the science. That is, they may deliberately choose to maintain or claim a state of ignorance about the technical issues in order to be able to define their position with respect to the issues involved. The ways in which this can be done are distinguished by three elements. The discursive form, that is the way in which the ignorance is expressed. The second element is the status that the form implies and finally there is the relation between science and the lay public that is indicated in the other two.

One discursive form is termed the non-scientific mind; this implies that the individual claims to have a status such that he or she cannot understand the issues involved, implying a relationship to science that is subservient and dependant. In the discursive form described by "not my job" the implication is that the division of labour is such that the individual does not need to make the decisions because there are those whose job it is to do so. The relationship here is one of coexistence and cooperation with a distinct demarcation of skills. These positions are interesting in that they mirror the deficit model since they implicitly lay the blame on science if any thing goes wrong, either because the individual is not capable of understanding the issues or because there is a clear division of labour or responsibility.

The third position is one in which the science is seen as missing the point because it does not deal with the germane issue or is concerned with some minutiae that can obscure the larger issues. The discursive form here is described as "not interested/relevant" and has the implication that a deliberate choice has been made to ignore the science because it has in some way missed the point or the nature of the matter in hand is that it is not scientific. An interesting example of this construction of ignorance is shown in the quotation attributed to Renee Elliot the founder of Planet Organic (organicfood.co.uk, 2001), my emphasis:

There's no Question! Organic food is grown and stored without the use of artificial pesticides and fertilisers. The fact that residues remain on conventional foods and are consumed by us over decades, accumulating in our fatty tissues is well documented. ***Ignoring this danger to health completely, as there has been no comparative long-term research about the difference between the long-term effects of consuming conventional versus organic foods, it is still obvious that organic food is better for you.***

If you look at beef and dairy as a category, organic food contains no antibiotic residues, growth promoters or BSE. ...

What matters to the author here seems to be not the fact that there is no scientific evidence, as explicitly stated, but the self-evident truth (for the writer) that the passage proclaims. The article goes on to list a number of the claimed contents of non-organic food and then asserts:

At best, these additives are unnecessary and annoying to those who question their use and usefulness. At worst, they are possible carcinogens and could be causing damage that no one has bothered to study.

It is easy and tempting for those scientifically trained to be dismissive and contemptuous of this kind of argument. But the point is that this is not actually a scientific discourse although it is apparently structured as such, it is something that is genuinely and deeply held belief by the author and is part of her world picture. It is more akin to religious belief or a belief in magic than science. Since she feels the statements to be self evident, and implicitly, that science is not needed to support them, arguments based on scientific lines are not going to make much difference to her and those who support her views. However it is interesting to note that even whilst the text contains dismissals of science there is still the use of technical terms such as "carcinogens" and an appeal to evidence, presumably scientific, about accumulation in fatty tissues. Ignorance therefore is not total but partial and selective. Science is not seen as indivisible but something from which one may select what is agreeable and dismiss what is not.

1.5 Words, Words, Words

Another example of the power of words arises from an anecdote. A friend found out of date organic baby food for sale in a whole food shop. When she challenged an assistant about this and pointed out the risks to health the response was "It's OK, it doesn't spoil because it's organic". Certain words have taken on new usages that carry very strong connotations with them. The anecdote and the passages quoted above imply special qualities for organic food that perhaps have more to do with its ritual qualities than its material ones. The way in which organic food is grown does not, of course, protect it from microbial spoilage, but the fact that it is grown organically seems to imply for some people that the process imbues the food with special properties rather akin to food that has been ritually blessed or has magic properties. There seems to be a vocabulary that has developed an alternative set of usages for particular groups. When used this can label products in a manner that either elevates them or stigmatises them, we might say sanctifies or damns them. To scientific experts these words have a different set of implications and connotations. Much misunderstanding can therefore be caused by the use of these words with variable usages attached to them. Some examples are given below.

Table 1.1 The usage of words with both scientific and popular connotations

Scientific Usage	Word	Alternative Usage
Relating to all compounds, whether naturally occurring or synthetic	Chemical	Artificial, factory produced, not natural, potentially dangerous
A phenomenon	Natural	Good, honest, healthy, clean, simple, safe
Related to the chemistry of carbon compounds	Organic	Pure, unadulterated, non-chemical
Appertaining to the study of heritability	Genetic	Dangerous, relating to inherited disease, unnatural manipulation
Concerning the emission of energy	Radiation	Appertaining to nuclear process, dangerous, insidious

This list is not exhaustive and cannot be; different communities will have different sets of words with different connotations. The point is that using a particular word to describe something can automatically put it into some category that will effectively ensure the end of debate. A scientist, trying to explain to a member of the lay public who is using the word "chemical", in the sense shown in the right hand column, that all matter is made up of chemicals and that chemicals can be naturally occurring as well as synthetic is not likely to get very far. This is not because the non-specialist necessarily disagrees with the scientist's usage, simply that the term is used to label a class of materials

regarded as undesirable. One of the reasons that science has developed its own specialised language, in which words are very tightly defined, is to eliminate ambiguity and ensure a common understanding of terms used. In a dynamic society, heavily influenced by technological development, it is inevitable that these tight definitions will become used in ways that extend the original scientific usage. The need is to recognise that this is happening and to be aware that an extended usage of a word is not irrational or ill informed but simply the consequence of language developing in time as it has always done.
It is a corollary of the failure of the deficit model that, if there are more ways of understanding and rationalising the world than the scientific, there are more ways of expressing ideas about the nature of the world than those limited by the tight definitions of the scientific language. However the situation is complicated by the fact that legitimisation of ideas tends to be within the scientific discourse, as a result of which the boundaries between the tight scientific use and the more general use often become blurred. This is particularly true of the word "risk" which can take on many meanings. The problems and nature of risk perception are dealt with elsewhere in this book. However the question of who is able to decide what constitutes a risk and to what extent a risk is quantifiable opens some very serious issues about the nature of science itself and these are considered in the next section.

1.6 Science and Reality

Within the deficit model of science communication is the notion that science itself is in some kind of privileged position; that scientific knowledge is in some sense superior. The idea that science has some special relationship to reality is strongly supported by the success of the scientific endeavour. The application of the scientific method has led to staggering achievements based on the ability to describe, predict and manipulate the material world in a quantitative manner. In spite of this demonstration of predictive power a considerable body of academic opinion has arisen which considers that science is not an observer-independent account of an external reality but that its output is dependent upon social and cultural networks and interactions. Generally this position may be referred to as one that considers science to be socially constructed (for readable accounts see Hacking 1999; Gergen 1999). This view has engendered considerable opposition among some members of the scientific community (Koertge 1998). This is not surprising since it is abundantly clear that science has been very successful. However it is important to recognise the rather more subtle ideas in the constructionist model: a set of equations describing some kind of behaviour in the natural or physical world may have predictive ability, but is the model on which they are based the only possible model of the world and does that model in some way depend on the social and cultural circumstances in which it was created?

Gergen (1999) sets out four "working assumptions" of social construction:
1. "The terms by which we understand the world and our self are neither required nor demanded by 'what there is' ".
By this he implies that there may be an infinite number of ways in which any state of affairs may be described and, in principal, none of these is better than any other. Indeed if reality exists it does not necessarily have anything to do with the way we behave in the world. This implies that science does not necessarily have any connection to reality *a priori.*
2. "Our modes of description, explanation and/or representation are derived from relationship".
Ultimately all that we can say about the world must be expressed in some form of representation be it spoken, diagrammatical or mathematical. Language does not arise autonomously within an individual; it arises, and derives meaning, from within a social context of exchange between human beings. Thus our descriptions of the world depend on relationships between individuals within societies
3. "As we describe, explain or otherwise represent, so do we fashion our future".
Since meaning and our behaviour in the world derive from relationships then such things as sustaining traditions, our future actions, the very nature of the world as we view it, are maintained or changed by our relationships. Thus in the way our interactions change and are changed by our representations, we recreate our world.
4. "Reflection on our forms of understanding is vital to our future well being".
The recognition that there are no givens, nothing is inevitable and that values arise because of social interactions is essentially liberating. Nothing *must* be. In this way a "generative dialogue" may be developed which can lead to common ground since nothing has to be right or wrong in any absolute sense.

As Gergen points out, not all social constructionists would agree with all these points. Since by the very nature of the constructionist view there can be no last word or appeal to authority there is no reason to assume that these views will not change. Certainly many scientists would find them very hard to accept, however they do give a lens through which to look at science and the scientific process that recognises that whatever scientists may feel about the special place of scientific knowledge, this view of epistemic privilege is not universally shared. In particular the idea of social construction allows a way of looking at the problems of prediction and risk in new areas of science.

The first problem when looking through this lens is how to relate its vision of all being possible with the common human experience of reality: aeroplanes fly independently of what we think; inert objects do not stay in mid air if dropped; people die. NK Hayles (Hayles 1995) has suggested a pragmatic way forward based on what she calls constraints. She assumes that there is in some sense an "out there" which she calls the flux. We interact with this flux through physical, biological and cognitive interactions. These, together with the social interactions discussed above, condition the way we see the world. However it is true that, however we see the world, it is constrained by our physical and biological interactions with the flux. Although we cannot achieve an "objective view" of the world and such a concept may not be meaningful, we do recognise that the world constrains us. Hayles illustrates the nature of constraints by considering gravity: our concept of the nature of gravity has changed radically from Newton to Einstein; we have moved from a concept of gravitational force in a universe in which there is an absolute frame of reference, to the concept of curved space-time in a universe where there is no absolute frame of reference. The universes these theories describe are different. They cannot both be "true". However they do both describe common human experience, which is that in the normal course of events a body will not remain in the air but will fall to the ground. This is one of the constraints of the flux. Thus even if gravity is a constructed concept the fact that things fall down is not. Rosa (1998) gives the example of celestial phenomena, particularly the sun. He points out that over the range of human history, and across the world the sun has always appeared to behave in the same predictable way. This and other considerations lead him to suggest the concept of "grounded reality" which he relates to high "ostensibility and repeatability".

An interesting case within these concepts is the second law of thermodynamics. A statement of the second law given by Partington (1950) clearly shows its origins in the nineteenth century interest in heat engines.

The operation of a cyclic process which produces no other effects than the abstraction of heat from a reservoir and the raising of a weight is impossible.

Hayles argues that this should not be considered as an example of a constraint, because, although it may be considered as an empirical generalisation, it is an explicit formulation and "a constraint that is expressed is a representation not a constraint" (Hayles 1995, p. 54). This may be the case with Hayles' definition but in fact the second law is a useful formulation of empirical experience as are many nineteenth century laws of physics and chemistry. It has been repeatedly tested and verified. As such, the second law is a good example of what I shall call empirical history. By empirical history I mean those things that many human beings over a long period of time and in many diverse places have been able to observe. It corresponds closely to the observables of science but does not imply any association or disassociation with an external reality. The problems of reality are deep philosophical ones: what can be known and what exists are ancient problems that have been debated by philosophers for millennia. However, interesting as they are, they are not relevant to the current question as we are concerned here with the nature of common human experience, whether or not that may be said to be in any sense "real". For it is common human experience of empirical history that forms our perception of the framework of the physical and biological world within which we move.

In principal, empirical history is not a predictor of the future because there is no way of being absolutely sure that the sun will rise tomorrow or that things left unsupported in mid air will not suddenly start to stay there, but it is the nearest thing that human beings have to certainty and the best predictor we have as to what will happen in the future. It is thus a good starting point for the consideration of the capabilities and limits of the scientific approach to risk.

1.7 Science and Risk

Empirical history is a measure of human experience reproducible over time and in different cultural environments independent of theory and explanation. If we wish to move from records of the past to predicting the future some form of induction is necessary. The tendency of things to fall down and the path of the sun through the sky are well attested. These are cases where the induction about the path of the sun in the sky and the tendency of things to fall down tomorrow may be considered to be strong. The induction that people will die tomorrow is a strong one, however exactly which people will die and under exactly what circumstance is much more problematic. Empirical history is still a guide; statistics do not allow the prediction of exactly who will die but they do allow predictions of the likely numbers to die overall, and further analysis will reveal in what age range and geographical locations these deaths are likely to take place. Empirical history thus allows the generation of a probability map. Insurance companies are able to use such statistics to offer life insurance; they cannot be sure which of their clients will die but they can be sure of the odds and can thus adjust premiums accordingly. The empirical history that insurance companies survive suggests that such a strategy is valid. Knight (1957) has formulated this kind of prediction as "insurable risk". The key factor in this concept is quantifiable prediction based on empirical history and repeated testing (Belton 2001). However the move from some kind of absolute prediction like the sun rising to a probabilistic prediction represents a weakening of the strength of the induction.

Clearly if insurance companies did not update their predictions as new evidence came in they would still be offering insurance premiums based on nineteenth century death rates rather than current ones. This makes them very vulnerable to any competitor who uses the most up to date data and offers much lower premiums as a result. The empirical history in this case shows that there has been a variation over time and the assumption being made is that the nature of the variation will be continued into the future. If the trend curve has a constant shape then the extrapolation is a good one. If this is not the case then the prediction will be wrong. Clearly here, the induction is getting weaker.

Finally there is the problem of prediction where there is no empirical history. This is the area where most new technology exists. The World Wide Web is in a state of dynamic flux on a global scale. What the effects of the web on society will be in 2060 are impossible to predict and even what will happen in the next five years is very hard to say because there is not yet sufficient empirical history. The only way to deal with the situation is to imagine possible worlds, that is to say, to think about things that might happen. The imagination of possible worlds is a process of creation which, like creation in all spheres of activity, derives from the social, cultural and psychological environment in which the creator works. Of course, the creative process is not independent of empirical history, it will tend to be constrained by it. A possible world in which the sun never rose is not one that is going to be likely to accord with future reality and will be a weaker induction than an imagined world in which the sun behaves as it always has. Nevertheless, even given the constraints of empirical history, the number of possible worlds is still infinite and the imaginative process, which creates them, falls well within the sphere of social construction. Imagining possible worlds is something that scientists do all the time, the creation of a theory or hypothesis is just such an act of creation. The imagined world is then tested against empirical history, as generated in the laboratory, and accepted or rejected. This process is not a smooth one, and is coloured by many non-scientific interactions, but the continued process of testing over time tends to remove the poorly predictive imagined worlds and leave the better predictors intact.

So far the problem of risk has not been dealt with. This is because risk, as discussed in other chapters of this book, and elsewhere (see for example Adams 1995a; Lofstedt and Frewer 1998; Belton 2001) cannot be quantified in any straightforward way since it combines both an element of prediction and an element of concern about outcomes. Thus John Ashworth (1997), talking about quantitative risk assessment writes:

... this 'technocratic' – as it is often called – view of risk assessment neglects the common observation that different people will judge the same risk in different ways, according to their personal circumstances and the context within which the risk is presented ...

Science cannot predict risk but what it can hope to do is say something about possible outcomes of events based on empirical history. Adams (1995b) has pointed out that risk prediction is reflexive. The identification of a risk and any attempt to quantify the likelihood of adverse events associated with it will in itself result in increased awareness of the risk and hence possible avoidance activity, this will then tend to decrease the number of adverse outcomes. The implication of this analysis seems to be that the notion of objective values for possible outcomes of events is therefore invalid, since merely noticing that a risk exists will cause avoidance behaviour and thus change the likelihood of possible outcomes. This view is predicated on the assumption that human behaviour is completely unpredictable. However, this is not the case, since there will be an empirical history of human behaviour in a variety of circumstances, and there are methods for taking such effects into account (see for example Burns et al. 1998). It is certainly true that perfect and exact quantification is not possible but in the real world where real decisions have to be taken, imperfect data is better than none at all.

What can be said and with what confidence depends on the quality of the data upon which prediction is based and where the prediction comes on the scale from empirical history to social construction. The relationship is shown diagrammatically in Figure 1.

There are no discrete barriers between the various elements in the scale; there is a continuous variation. This is because with the passage of time and repeated observation empirical history accumulates. Weak induction will either be rejected as not corresponding to reproducible reality or will gradually strengthen and become stronger. The validity of empirical history itself is not

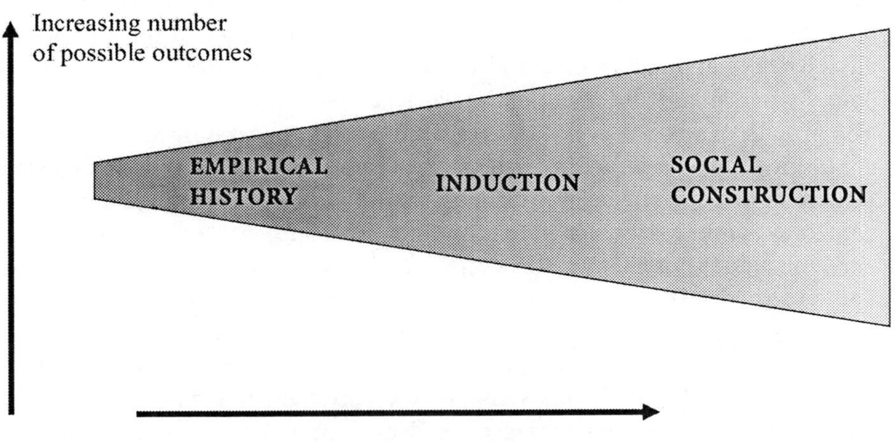

Figure 1.1 The continuum from empirical history to social construction

immutable, things change. Currently world weather patterns are changing and the old history of the seasons is unreliable. In this case the old empirical history is no longer valid for induction and instead new data with less validation must be used. Induction is therefore becoming weaker and some of the predictions will have strong elements of social construction. The message here is that the use of empirical history as guide requires eternal and critical vigilance to ensure that it does not itself become social construction.

1.8 Conclusions

Science in the postmodern world inhabits a very difficult environment. The very success of the scientific endeavour and the rate at which new science is turned in to new products in the market place has led to a situation in which science and scientific expertise are playing larger and larger political roles. In the process science has become the object of increasing scrutiny some of the elements of which have been summarised above. The first element of that scrutiny is the subject matter of science itself. As has been pointed out in this chapter, in the case of food the idea of food science itself contains something of a contradiction since the concept of food and its associations run well beyond the limits of science. Some care is needed here because it may be argued that food science is an applied science and the problem arises because of the broad definition chosen at the start of the chapter which combines science and technology under a single heading. However governments do not fund science because of its remoteness from application; science is funded because governments and commercial enterprises recognise the value of science in wealth creation and the support of public policy. There is thus a real sense in which all science is applied science and the gap between the esoteric activities of "pure" science and the commercial world is becoming smaller and smaller. The growth of science parks and research parks around universities and the continuing encouragement of scientists to become more entrepreneurial is a testament to these effects. It is difficult to find an area of science that is not in some way some closely related to economic, environmental or health concerns and which does not have an effect on the public and hence public perceptions. Some of the most active areas of science: molecular biology, environmental science, artificial intelligence, are at the forefront of public concern and commercial and political interest. All of these areas are loaded with extra-scientific connotations. Food, in this sense, is not a special subject and is typical of many other areas of science. Thus the distinction between science and technology is not a very useful one for clarifying the origins of the problems that science and technology now face. The general point to be made is that science does not operate in a vacuum and scientists must be aware that all of what they do exists in both a scientific and a non-scientific context. The failure to appreciate this and assume that only the scientific context is valid or real may well lead to confusion and hostility.

The confusion is further compounded by the assumption of the deficit model of science communication. The reality, as the chapters in this book and many other works show, is much more complicated. The continued use of terms such as "the public understanding of science" perpetuates the idea of the deficit model and should be abandoned in favour of a terminology that recognises the complexities of communication between scientists and the lay public about scientific ideas and recognises too that the issue is not about a one way communication process but about the development of a two way dialogue. Within this dialogue it will always be true that, whatever the intentions of scientific experts, their views and ideas will not be interpreted outside of a social and cultural framework, and the responses will be coloured by the relationships between the different sides of the debate. This perspective, and the need for dialogue rather than instruction, does not of course mean that scientific rigour should be relaxed, since this is exactly what the scientific disciplines should bring to any debate. Clear thinking and a clear statement of what is known, what is not known and what is uncertain is of enormous value. But the scientific input is not all there is, it is a component of what there is and its acceptance or not will depend in part on the style of delivery and on the perceived interest of the deliverer.

What is delivered and what counts as science can be problematic. If science is used on a pick and mix basis as a means of supporting one vested interest or another, its credibility is diminished. Neither can science be used in a partial way to support some position and then discarded or disregarded when either the data are not available or the results happen to be inconvenient. As was shown in the quotations about organic food, the scientific discourse is misused in this way. If genuine dialogue is to be constructed there is a reciprocal responsibility on scientists and lay public; both must recognise the limitations and the strength of science, which resides in its application of rigorous thinking and objective analysis of all the data, and both must recognise and legitimise other modes of thought about the issues.

Science is a dialectical process and at any point in time there will be a large number of unresolved problems. However, as time goes on empirical history is accumulated and what was undecided and uncertain becomes much more reliable. In the nineteenth century the possibility of heavier than air flight was uncertain, at the turn of the century it was shown to be possible but risky and poorly reproducible. Now flight is commonplace and aeroplanes fly with high levels of predictability. This does not mean that sometimes they cease flying unpredictably but on the whole the empirical history shows them to be reliable.

The challenge is to be absolutely clear where there is reliable empirical history and where there is not. By using empirical history as a criterion the epistemological problems raised by the ideas of social construction may be avoided. In essence the exact nature of scientific knowledge does not matter for practical purposes since how we actually live as human beings is to rely on the assumption that what has happened reproducibly in the past will continue to do so in the future. This does not mean that the unexpected does not happen or that things do not change. It is part if the human condition to recognise change and respond to it, however change can only be recognised by comparison to expectation, if there were no empirical history we would not be able to recognise change.

Where new technology is concerned there is often very little or no empirical history. There may be much science and development of the technology but there will have been no history of the large-scale deployment of the technology. Therefore reliable prediction of the consequences of the deployment of the technology is not possible; there may be educated guesses, there may be prediction by analogy or simply hope dressed up as something else. But it should be quite clear that predictions, where there is no empirical history, are the creation of imagined worlds and as subject to all the social, cultural and political influences as any other creative act. It is vital in these circumstances that science does not claim some special privileged knowledge and that it is quite clear about its limitations. This is not an argument against new technology; the use of technology has benefited mankind enormously. It is however an argument for the recognition of the fact that the decisions about the safe deployment of such technology cannot be justified solely on the basis of science.

Throughout this chapter I have referred to the limits of science and the areas where it cannot be applied. This is not to belittle science. It is among the greatest of human achievements, but its very success has led to a situation in which it has appeared to become the arbiter of truth and the only legitimate source of knowledge. This has led to great tensions because science is only one of the ways of seeing the world not the only way. Other world views have become buried and unacknowledged. They have not ceased to exist however and their presence has manifested itself in private, and increasingly public, rejection of science. By recognising and legitimising other views of the world constructive dialogue may be entered into which will not diminish or ignore science but give it its proper due as a powerful and useful way of informing action.

References

Adams J (1995a) Risk. University College London Press, London

Adams J (1995b) ibid. p 14

Ames BN, Gold LS (1999) Pollution, pesticides and cancer misconceptions. In : Morris J, Bate R (eds) Fearing food. Butterworth-Heineman, Oxford pp 19–37

Appleyard B (1993) Understanding the present. Pan Books, London

Ashworth J (1997) Science, Policy and Risk. In: Science, policy and risk, Royal Society, London, p 3

Belton PS (2001) Chance, risk, uncertainty and food. Trends in Food Sci Tech 12: 31–35

Burns WJ, Slovic P, Kasperson RE, Kasperson JX, Renn O, Emani S (1998) Incorporating Structural models into risk research on the social amplification of risk. In: Lofstedt R, Frewer L (eds) Risk and modern society. Earthscan, London, pp 163–180

Caplan P (1997) Food, health and identity. Routledge, London

Durant J (1997) Scientific truth and political reality. In: Science, policy and risk, Royal Society, London, p 48

Eagleton T (1998) Edible ecriture. In: Griffiths S, Wallace J (eds) Consuming passions. Mandolin, London, pp 203–208

Gergen KJ (1999) Social construction. Sage, London, pp 46–61

Greenpeace (2001): http://www.greenpeace.org.uk / contentlookup.cfm?CFID=19583&CFTOKEN=5941710&ucidparam=20011109134743

Hacking I (1999) The social construction of what? Harvard University Press, Cambridge, Mass pp 63–99

Hayles NK (1995) Searching for the common ground. In: Soulé ME, Lease G (eds) Reinventing nature? Island Press, Washinton DC, pp 47–63

Irwin A, Dale A, Smith D (1996) Science and hells kitchen: the local understanding of hazard issues. In: Irwin A, Wynne B (eds) Misunderstanding science? Cambridge University Press, Cambridge, pp 1–17

Irwin A and Wynne B (1996a) Introduction. In: Irwin A and Wynne B (eds) Misunderstanding science? Cambridge University Press, Cambridge, pp 1–17

Irwin A and Wynne B (1996b) Conclusions ibid p 215

Knight F (1957) Risk, uncertainty and profit. London School of Economics and Political Science, reprints series no 16. London

Koertge N (1998) A house built on sand. Oxford University Press, Oxford

Lofstedt R, Frewer L (1998) Risk and modern society. Earthscan, London,

Michael M (1996) Ignoring science. In: Irwin A, Wynne B (eds) Misunderstanding science? Cambridge University Press, Cambridge, pp 1–17

Morris J (1999a) Introduction. In: Morris J, Bate R (eds) Fearing food. Butterworth-Heineman, Oxford, p xvii

Morris J (1999a) ibid p xvi

Morris J, Bate R (1999) Fearing food. Butterworth-Heineman, Oxford organicfood.co.uk (2001):http://www.organicfood.co.uk/whybetterforyou.htm

Partington JR (1950) Thermodynamics. Constable, London, pp 26–27

Rosa EA (1998) Metatheoretical foundations for post-normal risk. J Risk Res 1: 15–44

Soulé ME, Lease G (1995) Reinventing nature? Island Press, Washington DC, USA

Yearley S (1966) In: Irwin A and Wynne B (eds) Misunderstanding Science? Cambridge University Press, Cambridge, pp 172–190

Food and Culture

2

ANNE MURCOTT

2.1 Introduction

Talk of food and culture extends well beyond the boundaries of arcane university research projects in social anthropology and sociology. Examples can be found with little effort – the following (with the emphasis added) were noted in passing while obliged to concentrate on something quite other:

A book review by Michael Bateman in one of the British broadsheet newspapers reports that the author, a top London chef is 'somewhat surprised' after a childhood diet of baked beans on toast 'to find that in 20 years he has ridden out a revolution in British food, even played a part in **shaping a new food culture**'. (The Independent on Sunday 8 July 2001)

The same idea – and the same periodicity – was pressed into service earlier in that year, when the same newspaper carried an editorial urging a far longer perspective on the newly reported outbreak of foot-and-mouth disease. The background to their urging noted the promotion by successive governments of 'the idea that mass-produced cheap food is desirable. The more meat that has been produced, the cheaper it has become.' Even the growth in vegetarianism 'has not stopped the rise in demand for cheap meat.' All of which, illustrates that '**(t)he culture of food and diet** has changed dramatically over the past 20 years.' (The Independent on Sunday 25 February 2001)

Introducing a radio discussion of the disappearance of regional variation in diet and daily menus such that it is no longer the case that what people ate in Devon is different from what is eaten in Derbyshire, the veteran food writer, journalist and campaigner, Derek Cooper referred once again to the 'loss of our **cooking culture**'. (BBC Radio 4 12 August 2001)

The makers and world-wide distributors of shoyu have set up the **Kikkoman Institute for International Food Culture** both a visitors' centre in Japan and website (http://kiifc.kikkoman.co.jp/english/)

Contributions from within the academy, but in disciplines other than social anthropology and sociology, are also found to mention food and culture. Nutrition itself is an obvious example, with some attention directed toward the social and cultural pioneered in Britain by John Yudkin (Yudkin and

McKenzie 1964) and continued there and elsewhere by, among others, Paul Fieldhouse (1986) in North America and Pat Crotty (1988) in Australia. New students to psychology now find themselves introduced to 'cultural determinants of eating and drinking' (Booth 1994, p. 70).

Certainly, substantiating what is mere impression illustrated haphazardly above is a further exercise. Checking how widespread such talk is, detecting whether it is relatively new, asking whether it is confined to the English language, and more, are separate questions requiring suitably systematic enquiry. Whether it is so confined or no, such further checking lies well beyond the scope of the present discussion. Yet, despite being just an impression it remains strong, and, as will be noted in passing, suggests that various kinds of older boundaries now seem to have become permeable. It is not only that the Kikkoman example represents a phenomenon that remains to be well named – simultaneously advertising, education and entertainment. It is also that (de)fences both within the academy (where increasingly multi- or inter-disciplinary work is urged) and those distinguishing it from worlds beyond, are either being stormed, lamented or simply ignored. The new, prize-winning journal *Gastronomica* (http://www.gastronomica.org) that is so ably edited by Darra Goldstein, represents the latter particularly well, both in the composition of its advisory board never mind that its subtitle, *The Journal of Food and Culture*, once again juxtaposes the concerns of this chapter.

So, tackling food and culture in the present context involves pausing to consider in a quite self-conscious fashion not just *what* is being talked about but the *way* it is being talked about – insofar as these are in any case distinguishable. It also means noting, in passing, that culture is a word which, on its own, enjoys seemingly widening use in public affairs, social commentary, marketing, public relations, business management and journalism, with reference to all manner of concerns – a 'blame culture' a 'culture of violence' a 'culture of secrecy', 'consumer culture', 'celebrity culture' or the need to 'change the culture' of an organisation. Very likely the reader is easily able to recollect the use of all these, never mind being able to add to the list.

Presumably usage of this kind serves the purpose of whatever context in which it is adopted sufficiently satisfactorily, or it would be abandoned and/or noisily challenged. There is, however, a case for reflection. What does all this seemingly newly minted talk of 'culture' and food – whether it be new trends in food production, eating away from home, novelties contrived in smart restaurant kitchens, or the reported popularity of ingredients originating in Thailand, an island in the Caribbean or the Australian bush – do for us that other expressions cannot? Has it perhaps become a catch-all, a fashionable synonym for 'ethos' or maybe 'lifestyle' that adds nothing in particular to our understanding? Perhaps it is no more than a useful noun where one is needed to get in reasonably good grammatical order to the end of a sentence. It is not that such talk is somehow idle or that it necessarily presents any difficulty or is in any way wrong. The point is, rather, that it may turn out to be too simple, drained of significance, to pose much of a problem or disagreement. Such simplicity can at times be especially useful, but may also thereby fail to make any telling contribution. Accordingly, one purpose of this chapter is to provide a basis for future consideration as to whether paying attention to some of the work (primarily in social anthropology and sociology) of the last 20 years or so might offer a grasp permitting more extensive and nuanced appreciation of food and culture. In the process, despite the complaint that food has been neglected in both the academy in general and in these two social sciences in particular, it may be noted that the literature on food and culture includes contributions from internationally very distinguished scholars in these disciplines. A further purpose is to indicate how thinking about food with reference to culture might pave the way for continuing examination of the relation between expert advice about, and different publics' views of, food and eating, safety, diet or nutrition.

What follows proceeds in episodic fashion, taking a simple form of variations on a theme. Addressing the editors' brief to write on 'food and culture' for a readership of non-social scientist, the chapter assumes little to no acquaintance with the academic disciplines for which culture is a central intellectual focus, analytic category or organising principle. The piece aims to be accessible, and veers well away from disciplinary technicalities – but liable, in consequence, to over-simplification. At the same time, its origins in a blend of social anthropology and sociology is evident in its style (the essay form), in its reach (anything from whole cuisines, century-long spans, to the small-scale of household meals) and in the overall stance adopted (distanced from cultural studies/cultural theory). So saying serves as the first 'lesson' for a reader completely new to this group of social sciences. Rather than consisting in cumulative findings, studies tend instead to follow one another not so much 'where the others leave off', but, 'better informed and better conceptualized, they plunge more deeply into the same things' (Geertz 1973, p. 25). In the process, endeavours get named and re-named, whole disciplines shift perspective, and segments sheer off or get subsumed elsewhere – and, introductions to the genre require extended introductions. The new reader, then, is well advised not to regard the present, or indeed any other *single* treatment of food and culture as definitive – and by the same token is well advised to be profoundly suspicious of any that claims to be so. Instead, readers are urged to reflect on the degrees of depth of whichever treatment falls into their hands.

That said, it is none the less possible to say a little about how some notion of culture ever become relevant to no matter which social science discipline's discussion of human food use. All would seem to have a shared – if infrequently stated – starting point that runs something like this:

Human beings are observed to be omnivorous. In parallel, any one individual is observed to be capable over their lifetime of changing what they eat. Human beings are also observed to be selective: they do not eat all that is available to them that is both non-poisonous and nutritious. In the face of extraordinary variety in eating habits within the species, biological and/or genetic explanations for such selectivity have to be regarded as incomplete. Thus, some realm peculiar to human beings has to be canvassed to achieve suitably comprehensive understanding – provisionally subsumed under the heading of 'culture'.

In this sense, it is not possible to describe the diet of the human species in the way that it is definable for other animals.

What counts as food, the very definition of food, is variable between different social groups and across different historical periods. Examples are numerous. Two are arbitrarily selected for mention. The suspicion with which the tomato was greeted on first introduction to sixteenth century England derived in part from being known to share membership of the same family as deadly nightshade, hence its being considered dangerous, with the result that it was first grown simply as an ornament (Wilson 1973, p. 309). The rejection of indigenous foods with whose use Aborigines were skilfully experienced, by the First Fleet on landing at Sydney Cove in 1788, who relied instead on insufficient provisions of their own familiar imported food, and failed attempts at agriculture, meant the initial settlement period became known as the 'hungry years' (Symons 1982; Cherikoff and Brand 1988).

The variety and selectivity in human eating coupled with the variability in what is counted as food, add up to the underlying theme on which the sections to follow represent variations, beginning with consideration of the dual biological/cultural character of human beings, then moving on to cultural understandings of 'the natural', cultural conceptions of meals, food and identity (nationality) and ending with food and class cultures. The selection is parochial, with examples confined to Britain, and is made with a view to illustrating the high degree of variety in academic attention to food and culture. All the same it is inevitably arbitrary and is not intended as either a comprehensive or critical review. The sections can be read in any order – some readers may want to pick up one or two substantive examples before reading the second section which precedes all these briefly to discuss the term culture itself.

2.2 Food and Culture: Disciplines and Definitions

So far, this chapter has glossed over distinctions in usage of the term culture. Even when it is linked with food, the previous section ignored any disjunction between the chapter title's reference to 'food and culture' and the handful of examples presented at its beginning which use the particular phrases 'food culture' and 'cooking culture'. This section remedies the oversight. It does so not so much in order to clear the ground – a task well beyond the scope of this essay – but in order to make a start on identifying what is growing so vigorously upon it, with a view to achieving some purchase on the import of the analytic notion of 'culture' as used by academics (mainly, as already indicated, social anthropologists and sociologists). It should be noted that defining usages of culture may be beside an analyst's point, and frequently has to be left undiscussed.

Short of resort to the chronological, there is no obvious place to start that is not arbitrary, so, beginning with a recently published collection of readings pertinently entitled *Food and Culture* (Counihan and van Esterik 1997) is as good as any. Bringing together selections from major authors, the volume is promoted on its back cover as 'crossing many disciplinary boundaries... including... the perspectives of anthropology, history, psychology, philosophy and sociology' – all disciplines for which culture has been a main analytic concern. How do the editors – both university anthropologists working in the US – introduce their topic? What is their talk of food and culture?

Given their task, it is probably unremarkable that, by way of avoiding preempting individual chapters to come, they do not offer a definition of culture – any more than they offer a definition of food (which, as has already been noted, is itself variable within the human species). That they mention the word culture or its variants some 20 times in just 5 pages is hardly surprising. Much more remarkable is that they use it in at least 7 different ways, referring to:

(1) a nameable social group or conglomerate of social groups and/or their associated world-view, customs and mores that is/are distant from or unfamiliar to the commentator together with: 'the (book's) articles... offer(ing) diverse perspectives about other cultures'

(2) variations on human universal themes, that which distinguishes human beings: 'how women across cultures so often speak through food and appetite'

(3) variations within those universals: 'cultural images of masculinity and femininity'

(4) one among other modes of scholarship, schools of thought: 'theoretical approaches... semiotic, structuralist, materialist, and cultural'

(5) a political stance that is tolerant of difference, paralleled by policies that take account of difference: 'cultural acceptance of diverse (women's) bodies is one key step in female empowerment'; 'the development of culturally appropriate messages regarding optimal nutrition'

(6) one among other bases for social disadvantage: 'culturally and economically marginal people often suffer hunger and malnutrition...'

(7) a broad arena alongside others such as the nutritional, economic, or policy: 'nutritionists need to pay attention to sociocultural factors surrounding food'.

The inclusiveness of this list can be set alongside a mid-twentieth century International Encyclopaedia of the Social Sciences (Sills 1968) note of the 'charter distinction of the anthropological concept of culture' by Edward Tylor in his 1871 classic *Primitive Culture*:

Culture or Civilization, taken in its wide ethnographic sense, is that complex whole which includes knowledge, belief, art, morals, law, custom and any other capabilities and habits acquired by man as a member of society.

Reference, incidentally, to the term's wide, 'ethnographic sense', further underscores the inclusivity of the notion. It can do so by directing attention over and above a specific notion of culture that refers to the aesthetics and sensibility of that high culture associated with ('Western') social elites' self-conscious appreciation of the arts – theatre, literature, sculpture and so on – which has been valued above low, mass or popular culture, but which may now be valued equally with, or even, connoting an inverted snobbery, below it (for a discussion of reference to 'high' culture as a historically specific usage see Williams 1983, and for its extension in cultural studies see e. g. Brooker 1999; Kahn 1995).

The inclusive version of the term silently counterposes culture to all that is biological or genetic, introducing a distinction between nurture and nature – though, it needs to be noted, not automatically privileging the former over the latter.

So inclusive a definition, this kind of '*pot-au-feu*' as he aptly dubbed it, can lead to what the Princeton anthropologist Clifford Geertz described as 'a conceptual morass' (1973, p. 4). In a little exercise which prompted the above inspection of Counihan and van Esterik's usage, he counted 11 different definitions of culture in a text he describes as one of the 'better' introductions to anthropology. Instead, Geertz explains that the concept of culture he espouses himself is, above all, an analysis not, of 'an experimental science in search of law, but an interpretive one in search of meaning' (Geertz 1973, p. 5). Later, in the same set of essays, Geertz enlarges on his position declaring that, for him, culture

denotes an historically transmitted pattern of meanings embodied in symbols, a system of inherited conceptions expressed in symbolic forms by means of which men communicate, perpetuate and develop their knowledge about and attitudes toward life. (1973, p. 89)

Self-evidently the several terms he uses in his definition require explication (and no doubt he would nowadays amend 'men' to 'humankind' or some such). But for present purposes, an emphasis on meaning serves well as one of the main flags or beacons on which to keep an eye. Another is the stress on knowledge.

There are several connotations in Geertz' reference to knowledge. One is very broad. It does not concern a specific body of knowledge so much as that confident way of 'knowing about the world' (in both the sense of the world in general and their world in particular) which each one of us can deploy, no matter to which society or social group we belong. This is the knowledge which *every* human being (with the probable exception of those mysteries, feral children who survive Romulus and Remus fashion) starts to acquire from an impressively young age, comes to take for granted and, possibly most significantly of all, *forgets they had once to learn*. Work considered in what follows exemplifies this form of knowledge in respect of all kinds of food use. Another connotation referred to is the mode of apprehending the world developed among specialists in no matter what – the creative in some craft, the spiritual in a priesthood in a system of belief and observances, the forgotten or lowly in sewage work or grave-digging – as well, of course, in what in modern 'Westernised' society and discourse is named as science, its practitioners and its institutions. Analyses of the manner in which specialist knowledge is transmitted to novices, shows that it is not simply the technical know-how that is at stake, but also the deportment, demeanour and etiquette, the insider slang, the outlook on the clientele and/or the relevant laities in short, their 'trained', i. e. socialised knowledge of the world. And it is this emphasis on meaning and these connotations of knowledge in the use of the term culture which thread their way through the literature on food and culture.

Before proceeding to examine particular examples, one further point about the analytic usage of culture has to be made. This emphasises the dynamic and the historical. More than once through the second half of the twentieth century, social anthropology and sociology each came to be roundly criticised for entailing an empirically inaccurate image of their subject matter as static and unchanging. For understandable reasons – study takes its time, and holding the objects of investigation still in order to look at them makes the work more manageable – terminologies together with their referents tend to become unwarrantably fixed, culture being no exception. There are, though, other reasons. Critiques of colonialism and social anthropology's relation to it have entailed dismantling an implicit idea of 'us' and 'them' – where 'they' may well be exotic, noble and able to teach 'us' about the virtues of simplicity despite 'deserving' to learn from 'our' knowledge, science and technologies – where 'their' ways of life could be characterised as homogenous, 'pre-literate' and without history. Rebuking anthropologists for leaning towards the simplification and temporal selectivity of so doing, Clifford insists it be remembered that '"(C)ultures" do not hold still for their portraits' (1986, p. 10). His memorable phrase appropriately plays up the dynamic in the notion of culture, combined with the related critique of a-historic analyses, which together represent a further theme in the selection of literature on food and culture to which the remainder of this chapter is devoted.

Before turning to that work, a final comment is in order. The question of historical change, of the dynamic, the developmental, does not always receive the prominence and analytic attention it requires to make adequate sense of the evidence in social scientists' use of the term culture. Whether they are human geographers, sociologists or anthropologists, and especially when adopting the phrase 'food culture', it will continue to be important to guard against casual, if unintentional, neglect of its analytic import. As colleagues and I cautiously noted when eventually opting to use the phrase

> *'Culinary culture' is a shorthand term for the ensemble of attitudes and tastes people bring to cooking and eating. Since 'culture' is understood in sociology and anthropology to mean all that is 'learned, shared and transmitted' among groups of human beings from generation to generation, it is not surprising that the idea of culinary culture has been associated with research of an historical-sociological kind aimed at explaining how different social groups – especially different societies or nation-states – came to* ***develop*** *different tastes and attitudes over time.*
>
> (Mennell, Murcott and van Otterloo 1992, p. 20, emphasis in the original)

2.3 Food and Culture: Cooking, Classifying and Being Human

The place mats of a 1970s Edinburgh restaurant bore the following aphorism:

> *... I had found a perfect definition of human nature, as distinguished from the animal... My definition of Man is 'a cooking animal'.*

This quotation (the place mat also provided the reference, James Boswell's *The Journal of a Tour to the Hebrides with Samuel Johnson*) neatly summarises one approach to the analysis of food and culture. For it picks up on an implication in the distinction suggested above between animals on the one hand and, on the other, people reared, of course, in human society. Culture has to do with learning, depends on the human capacity for abstract thought and memory, the ability to handle symbols and to attribute meanings that – as far as is known – outstrips anything similar in the animal kingdom. This is held to be neatly encapsulated in the universal human practice of cooking food. It is human beings who deliberately apply heat to much of what they eat, a predominant mode of transforming items they name as raw into an irreversibly altered state they accept and name as ready to eat – whether it be roasted directly against a fire, or via a medium other than air, e. g. boiled in water or submerged in oil.

Claude Lévi-Strauss, the French anthropologist, attends to this matter in asking what it is that makes the human species distinctive (1964 [1970], 1966 [1965]). At the same time, he takes due account of the unavoidable fact that human beings are also animal, they have a dual character. The species to varying degrees closely resembles other animal species, with parallel biological requirements and bodily functions. Yet the human species is unlike animals by virtue of being human. It is in this way that humans are creatures of *both* 'Nature' and 'Culture'. Therein lies a tension, one that is paradoxically permanently irresolvable yet continuously demanding resolution.

Expression of this conundrum facing the species maybe found, Lévi-Strauss proposes, via a scheme he called the culinary triangle – part of an extended analysis which borrows from a branch of linguistics. It is built up from a pair of binary oppositions: transformed/not transformed (*elaboré/non elaboré* or marked/unmarked) on the vertical axis with Culture/Nature. Transformation may take place with or without the deliberate intervention of human hand, either by cooking – Culture – or by biological processes of putrefaction – Nature. This double opposition is put together to form

> *a triangular semantic field whose three points correspond respectively to the categories of the raw, the cooked and the rotted. It is clear that in respect to cooking the raw constitutes the unmarked pole, while the other two poles are strongly marked, but in different directions: indeed the cooked is a cultural transformation of the raw, whereas the rotted is a natural transformation. Underlying our original triangle, there is hence a double opposition between* ***elaborated/unelaborated*** *on the one hand and* ***culture/nature*** *on the other.* (Lévi-Strauss 1966, p. 937 emphasis in the original)

Cooking is thus revealed as peculiar to human culture, deliberate transformation of the raw material by human hand that marks it off from natural transformations of ripening and putrefaction. This scheme, Lévi-Strauss suggests, is just the start: he devises other more developed triangles and even indicates that other, more complex, geometric schemes are possible. In all, food use for Lévi-Strauss expresses something more than just eating: food is not simply 'good to eat' it is 'good to think'.

Perhaps unsurprisingly, these analyses of food and culture over the three decades or so since Lévi-Strauss first published his culinary triangle have generated a varied response. These range from brisk dismissal (e.g Mennell 1985) criticisms of insufficient attention to evidence (e.g. Farb and Armelagos 1980) – food for the Japanese is raw or uncooked food, transformed artfully by skilled human hand, but raw nonetheless (Bestor 1999, p. 223) – and scepticism as to reliance on binary oppositions (e.g. Douglas 1984; Goody 1982) alongside more sympathetic treatments (e.g. Leach 1970, 1976). In part, the more sceptical responses can be accounted for by the manner in which Lévi-Strauss' underlying concerns are antithetical to Anglo-Saxon social anthropology: his interest lies in identifying universal features of 'the human mind', regarding some attitudes to food as inbuilt to the human brain. All the same, there are those finding his work sufficiently thought provoking among English social anthropologists to have developed imaginative analyses of e. g. children's use of confectionery (James 1981) and 'alternative' dietary remedies and fashions for health- and whole-foods (Atkinson 1978, 1983) which rely directly on his insights.

Furthermore, his influence persists. Presenting recent results from a survey of Nordic eating patterns, Ekström and Fürst (2001) refer approvingly to his triangle, asserting that the art of cooking which attempts to resolve the ambiguity of simultaneous existence as of both nature and culture. They go on to claim that

> *(t)his is why cooking is a culinary activity that involves creativity. The raw material is thus transformed into something beyond the mere edible, into something palatable, pleasurable and enjoyable for nose, eyes and mouth*
>
> (Ekström and Fürst 2001, p. 214).

It might be added, however, that their rationale for slipping across into cataloguing positive sensory aspects of eating is not immediately obvious. Indeed, their assumption that eating is pleasurable may well be a product of their membership of a social group in which eating is valued in this fashion. Not all groups do so. Almost as an aside, Geertz records the way he finds that eating is regarded by Balinese people:

> *The Balinese revulsion against any behavior regarded as animal-like can hardly be overstressed. Babies are not allowed to crawl for that reason... The main puberty rite consists in filing the child's teeth so that they will not look like animal fangs. Not only defecation but eating is regarded as a disgusting, almost obscene activity, to be conducted hurriedly and privately, because of its association with animality.*
>
> (Geertz 1973, pp. 419–420)

In this case, the anthropologist finds that any transformation of raw ingredients by cooking appears to fail as a means of resolving the ambiguity of human existence. Either way, however, it is possible at a certain level to see how it is that resort to an idea of culture as the realm of attributing meaning to items people use, in this case foodstuffs and associated utensils and cooking techniques, as well as biologically essential activities deal in aspects of human identity acknowledging that we are animal but also distinct from, even more than animal.

Despite being venerable, if not currently somewhat unmodish, this approach to food and culture is one which continues to reverberate in the literature – with binary oppositions continuing to resurface (cf Warde 1997). Note that here culture is at issue in at least two if not three ways. First in opposition to nature, i. e. the non-human; then as the very realm via which the first is apprehended. And third, of course, it is analysed within the boundaries and discourse of later twentieth century French structuralist social anthropology, i. e. via a highly specialised mode of knowing.

2.4 Food, the Body and the Senses: Cultural Shapings of the Natural

Where in the previous section, one mode of apprehending culture was in opposition to nature, this section approaches 'food and culture' from a different angle. It entails considering the way that nature can be said to be 'fashioned', shaped or constructed by culture, attending, in other words, to cultural conceptions of the natural. Two illustrations are presented here: one centres on the body, the other on the senses.

2.5 Food and Cultural Conceptions of the Natural Body

Vegetarianism provides an example – or, more exactly, the late twentieth century version of vegetarianism that is closely allied to the wholefood movements of the UK, the US or Australia.

Espousal of this variety of vegetarianism and health-foods is commonly found to be couched in a vocabulary of ethics. It is an ethics that draws boundaries in different places, tending primarily to accord animals rights that are akin to those which are accorded to people. This in some versions extends to doing all possible, when selecting items for food that not only eschews animal sources, but avoids uprooting or killing plants, just taking seeds, fruits or leaves. In the process, fruits and seeds are held to be especially valuable, for they are held to be especially invigorating, vitality-giving, the parts of a plant that are most 'full of life'. At the heart of such ethics is a notion that the planet, along with its flora and fauna, is to be conserved by attempting to reverse the ravages of industrialisation and restore its earlier purity. Highly refined and processed foods are deemed artificial, thereby unacceptably remote from the purity of the natural, the whole and wholesome.

Lupton summarises much of the literature's interpretation of such findings:

> *...the processing and refining of foods serve to detract from their inherent 'goodness', rendering them 'non-foods'. The dominant appeal of health foods is their imputed ability to restore purity and wholesomeness, to retreat from the complexities of modern life to an idealized pastoral dream of the 'good life'.*
>
> (Lupton 1996, p. 89)

She goes on to note that food marketing has picked up on this form of a cultural conception of the 'natural'. An associated lexicon has been invented –

'healthy', 'real', 'fresh', 'natural' – exploiting alignments between binary oppositions of natural/artificial, unprocessed/processed, healthy/unhealthy. A further version of such an opposition would seem to have developed more recently. This is an overlay, which simplifies into a binary distinction creating an opposition between foodstuffs containing genetically modified ingredients – artificial, unnatural, unhealthy – and those which are produced by methods accorded organic accreditation – untouched, natural, healthy (although, in the absence, so far as is known, of any investigation of the matter, a small student dissertation could usefully confirm this impression).

Lupton, however, continues to consider a different cultural conception of the natural, one, which she points out, is ignored by the vegetarian/wholefood construction. This is the conception of the natural informed by the life sciences, by toxicology and food science. She invokes familiar instances: commercial food processing can protect rather than endanger health by removing naturally occurring toxins, offering improved preservation techniques and the like; home preparation and cooking may impair nutritional content to a greater degree than factory processing.

A parallel contrast between a conception of the natural remote from the formal sciences and one directly emanating from them can be illustrated from Pill and Stott's (Pill 1983; Pill and Stott 1982, 1985) very well known work on the understandings of diet and health in an urban, socio-economically homogeneous group of women in a lower income category. Women's understandings of the role of diet in health/illness divided into two: those dubbed 'Lifestyle' and 'Fatalist'. The former held a kind of free-will view that individual behaviour, over which one had a degree of control and even moral accountability, is relevant to aetiology. Among this group, assuring resistance to illness was possible by looking after oneself properly, in particular by keeping to a well advised diet. The latter, however, leaned towards a variety of predestination, emphasising characteristics deemed immutable, and beyond anyone's control to alter, whether for better of worse. Such characteristics included heredity and family susceptibility, individual susceptibility, the type of person you are (happy, nervous sort) or the temperament you were born with.

Both Lifestyle and Fatalist involve a notion of the body, indeed both involve a notion of what the body 'naturally' is, despite being diametrically opposed as to the degree of its mutability. Although not analysed in such terms, these findings represent cultural conceptions of nature, in this case represented in the human body and varying in respect of the degree of 'nurture' that could affect a given 'nature'. Whatever one thinks of either version of the cultural shaping of the natural, Pill and Stott's material also, of course, starts to expose what may lie behind an apparent mismatch between professional advice and readiness to accept and comply with that advice. Indeed it reveals a coherence and internal consistency in attitudes which, from another perspective could be judged irrational rejection or simple misapprehension of modern scientifically informed dietary advice.

2.6 Food and the Cultural Shaping of Taste

That even the sensory is not only physiologically identifiable but also culturally shaped was hinted at in the familiar examples provided in the opening passages of this chapter. What is meat to one is, in the old saying, poison to another, not necessarily literally lethally, but certainly metaphorically. Disgust can overtake any one of us: being presented with cocks' combs as adventurous ingredients of ultra-stylish haute-cuisine evoked classic symptoms – pallor, retreat from the stimulus, raised shoulders, and swallowing in attempts to stave off the gagging reflex – in diners at an elite twenty-first century Paris restaurant. What is nutritionally valuable can not only simply be conceptually abhorrent, but also provoke a well-described syndrome of revulsion. This is all of a piece with Lévi-Strauss' dictum that food is 'good to think', of the symbolic realm in which items are culturally classifiable as either food or non-food – or as in this case, culturally, anything but food.

Among the critics of Lévi-Strauss' analyses, one of the most eloquent is Sidney Mintz, social anthropologist still working at Johns Hopkins, though long-since officially retired. He does not disagree with the idea that food is good to think, so much as argue that by way of an approach to food and culture it is incomplete. Analyses such as those of Lévi-Strauss (and others such as Mary Douglas whose work will be considered later who adopt approaches parallel to his structuralism) are subject to the powerful challenge that they are a-historic. In other words they not only reify the notion of culture, but treat it as if it were some kind of static blueprint of one or other social group's worldview. Mintz argues that far from being static, it is subject to change, simply by virtue of the practical being analytically prior to the symbolic. The human capacity to attribute meaning may be constant, but the particular meaning attributed is not once and for all. Quite the contrary, it is amenable to subtle shifts in the process of human use of the items in question.

Mintz' best known exposition of this position is found in his extended treatment of a remarkable case of the cultural shaping of taste – the case of a preference in some nations (e.g. Britain) for sugar on so massive a scale that it becomes tempting to speak of a 'national sweet tooth' (Mintz 1985). Certainly he grants that there is evidence for a biological predisposition in the human to favour sweetness. But this cannot, he argues, adequately account for the well documented extent of the preference, the types of taste and sheer volume of contemporary consumption. In any case, as he tartly observes, no-one is being force-fed sugar. To identify the factors sufficiently powerful to explain so massive a demand, he turns to document the social and economic history of its production and consumption.

Mintz demonstrates that the meanings attributed to sugar changed over four centuries. Towards the end of the Middle Ages, European knowledge of sugar defined it not as a foodstuff but a medicine – at one stage, held to be proof against plague. Once embarked on its career as a food, sugar initially signified great wealth and privilege, the prerogative of monarchs, courts and aristocracies, shifting with increasing supply and supplier induced demand to a mass taste which, particularly in the form of treacle, became recognised as an everyday necessity, providing a substantial source of calories for the nineteenth century urban working class of industrialised Britain. Sugar is certainly 'good to think', but since the 'thoughts' varied markedly over the centuries, adequate analysis requires including the socio-economic forces that create the circumstances in which it gets to be used thereby creating new opportunities for the attribution of historically shifting meanings. Although widely and approvingly referenced, it would seem that as yet, others have yet to develop parallel analyses – although a preliminary version was recently attempted to reveal the manner in which 'craving' – a particularly florid example of the supposedly physiological/sensory – for chocolate reportedly greater among women than men in modern Britain can be shown to be culturally shaped (Gofton and Murcott 2001).

2.7 Food and Culture: Conceptions and Significance of Meals

This section doubles back a little, temporarily setting aside thinking about grand sweeps of history in order to focus on the small scale of eating as a mundane, an unexceptionable and more or less routine, day to day activity. The discussion begins by noting that it is an observation common to many perspectives in social anthropology and sociology that on the whole human affairs are remarkably orderly: they become disorganised from time to time, but they are not chaotic. In other words, it is possible to uncover regularities, conventions and rules, or norms, underpinning the manner in which the ordinary features of daily life are organised and in the way that people get through the day. In key respects the social analysis of meals depends on this observation.

One of the especially influential analysts is Mary Douglas, the London University professor of social anthropology, whose interest in the analysis of food and culture runs pretty much the whole length of her distinguished career (e.g. 1966, 1972, 1975, 1984). Where Lévi-Strauss insisted food is good to think, Douglas declared, in the title of one of her short articles, 'Beans means thinks' – an allusion which more middle-aged readers will recognise to an advertising slogan for a well known brand of tinned baked beans (1977). And, in just the same way that linguists can uncover regularities in language, so Douglas – in parallel with Lévi-Strauss – argues anthropologists can decipher arrangements for eating. In particular, she argues it is possible to derive rules for that type of food event that in English is known as a meal which, unlike Lévi-Strauss' analyses neither depend on binary oppositions nor aspire to describing human universals. On the contrary, Douglas insists, meals are to be deciphered in terms of themselves, paying detailed attention to the characteristics distinctive of a locally occurring instance. They are to be understood in relation to the social context in which they are found, not torn from it.

A notable example of her work was conducted some 30 years ago with one of her students, Michael Nicod, who used the classic social anthropological technique of participant observation, which, in modern England resolved into living with families as their lodger (Douglas and Nicod 1974). In order to examine what counted as a meal, Douglas and Nicod stepped aside from the conventional names to coin the (albeit slightly stiff) expression 'food event' to serve the purpose of recording any occasion when either food or drink was taken. This allowed them to place familiar occasions in an unfamiliar light to examine what is well-known but by the same token typically unnoticed. In contrast to the event colloquially known as a snack (which can be taken solo, at what can be readily recognised as odd – i. e. non-meal times – standing, sitting or walking, indoors or out et seq) meals are structured affairs in respect of time, place, companions, use of crockery and cutlery, and of course, menu.

Based on Nicod's observations of household meals, he and Douglas were able to derive a series of rules governing the three types of meals they identified, their timetabling and sequencing – daily and weekly – and their status and degree of flexibility, relative to one another. The main meal of the day is the most highly structured with rules for its component parts in terms of the sequence of courses, combinations of dishes and elements of each dish being equally strongly structured. With its two courses, the first, main course of hot, savoury food based on a three-fold structure of a protein source (meat or possibly fish or egg) as the centrepiece, a starch, typically potato and additional vegetables with liquid dressing – the cliché of English cuisine 'meat and two veg' accompanied by the often mocked gravy. The second course is sweet, also hot, and recapitulates the structure with a starch (sponge, pastry) a centrepiece of cooked fruit and again, a liquid dressing, as in apple pie and custard.

Later work acknowledged the debt to Douglas, adopting similar (but not identical) analytic modes of procedure and confirmed the strength of the rule-based arrangements. Notable examples showed the proprieties underlying the combination of dishes, form of presentation, taking etc. in South Wales (Murcott 1982, 1983a, 1983b) and across the country in north-east England (Charles and Kerr 1988). The South Wales work stressed that these arrangements, often colloquially described as a 'cooked dinner', constituted a conception of a proper meal as an *ideal*. Practice was recognised as departing from the ideal, e.g. for speed's sake, but as long as what was deemed a suitable quota of cooked dinners were provided through the week, with a more elaborate version on a Sunday, then this aspect of the household's health and welfare, and orderly family life were held to be assured. The production of such meals was shown to be of a piece with conventional divisions of domestic labour whereby women – typically cast as 'homemakers' – cooked meals as a domestic service to men – typically cast as 'breadwinners' – as part of the marital bargain, the 'give and take' in each partner's making their appropriate contribution to the household.

At the time, i.e. the 1970s and 80s, reports of the idea of 'meat and two veg' as a lynchpin of domestic British eating tended to be regarded as unexceptionable – even banal. Subsequently some researchers (e.g. Keane and Willetts 1995; Kemmer et al. 1998) (based on a not altogether accurate supposition that the South Wales work was report of actual, as distinct from idealised, eating patterns) have cast doubt on the continued relevance of such analysis, claiming that meals were no longer so clearly structured and that domestic divisions of labour were no longer so strongly gendered. While there is further work to be done to bring together this more recent research with marketing reports of dramatic changes in British menus, the diminished reliance on home-cooking, increased frequencies of eating out of the home, it, in the meantime, can be suggested that the researchers in question tend to approach the analysis of meals in an unduly narrow fashion. First, they appear to assume the analysis applies only to a certain example of a structured main meal – neglecting other national variants (for examples of the import by migrants from the Asian subcontinent and Italy, see Bush et al. 1998; Williams 1997).

Second, these researchers incline to neglect another dimension Douglas made clear. Efforts at improving health by advising changing dietary habits are bound to fail if the manner in which eating is structured is ignored. It is no good, she robustly pointed out, dieticians proposing that the balance between fat, sugar and fibre intake be adjusted in a professional nutritionally advised direction by replacing that second course of the main meal with a piece of fruit. Such a move is antipathetic to the meal's structure. It is not the case, however, that shifts cannot or do not occur. Changes – whether secular or policy driven – are, she argued, most likely to happen where the overall structure is weakest. So here, once again (although Douglas and Nicod did not express it in these terms) is an instance of a mismatch between the orientation to dietary intake among experts and the daily eating arrangements, the use, experience and cultural practice of members of the public.

Along with the changes since the 1970s and 80s work to the use of cook-chill meals, more highly processed foods, the marketing reports of the popularity of dishes such as chicken tikka, 'fast' foods such as burgers or pizza – especially amongst children – is heightened concern about the low intake – again, especially among children – of fruit and vegetables. In addition, there is also a suggestion that learning about the idea of a cooked dinner as well as learning to appreciate partaking of it may yet be shown to co-exist with, rather than having been superseded by, these trends. As part of development work devised to support the improvement of school food policies, small-scale pilot investigations were recently undertaken of the diets of primary schoolchildren in an underprivileged area of Birmingham (Kyle 2001). Reflecting national data, soft drinks, cakes and chocolate were found to figure prominently in the diets of these children. Citrus fruits were eaten by approximately only a quarter of the children, and

Green vegetables such as cabbage, greens and broccoli were also unpopular, being consumed by 39% of boys and 44% of girls during the seven-day period.

(Kyle 2001, p. 4)

The study extended beyond collecting data on dietary intake to include some sense of the children's (as well as parents' and teachers') attitudes to food. Interviews with groups of children confirmed a predictable list of 'favourite' foods or dishes: pizza, hot dogs, kebab meat, burgers, shepherds pie, meat pie and spaghetti bolognaise.

Rather more unexpected was the way that all the children interviewed switched from talk of dishes to talk of meals to mention one meal in particular. Apparently universally enjoyed, this was the cooked dinner, sometimes referred to as 'a dinner'. Asked to describe a dinner, children talked of a meal identical to the South Wales version of 20 years earlier, i.e. meat, potatoes, at least one additional vegetable, and gravy. Indeed it is striking how close the accounts Kyle collected are to the data presented in the earlier studies:

Children in this part of Birmingham in 2001 defined what they meant by a cooked dinner in precise terms both by referring to the constituents of such a meal, and also to when, and with whom it was eaten. Data gathered opportunistically during discussions about food at a health conference for 13 and 14 year old pupils from schools in (the area) seemed to confirm that children in this area understand the symbolic significance of the 'cooked dinner'.

Most children interviewed said that they had a cooked dinner every Sunday, but a few also reported eating this type of meal on one other occasion during the week, an occurrence perceived as 'being very lucky'. Children drew a clear distinction between what they meant by the term 'cooked dinner' and other types of meal. Although meals eaten during the week were obviously cooked, they were not accorded the same status. Unlike meals eaten during the week, the cooked dinner was usually eaten with other members of the family. In fact many children reported that they ate this meal with their grandparents and other members of their extended family.

(Kyle 2001, p. 15)

Children's reports of their preferences and eating patterns were confirmed in interviews with parents, as well as teachers. Throughout, the idea of meals and the manner in which they are structured is apparent.
As if the children in her study were well tutored in Mary Douglas' approaches to the analysis of meals, possibly the most intriguing is Kyle's observation that

> *Children have a strong sense of the appropriateness of food combinations. It was clear from this small pilot study that the apotheosis of a meal was the 'cooked dinner'. This was the meal at which vegetables were eaten without complaint. However vegetables eaten as part of this particular meal do not turn the child into a vegetable eater in other circumstances. Cooked vegetables were not seen as an appropriate accompaniment to pizza or burgers, even if this combination might be viewed as nutritionally adequate.*
>
> (Kyle 2001, p. 30)

Here, then are what might be called good 'cultural' reasons for a less than well advised diet. It is in terms such as these that it is possible to make sense of what from other points of view are experienced as a gap between expert advice and the behaviour of individual members of a social group.

2.8 Food and Culture: Nationality, Ethnicity and Identity

As was indicated right at the beginning of this chapter, there is repeated report of variability between different social groups as to what does and does not get to be defined as food. This extends to encompass variations in conceptions of types of food event. While there may be common elements at a certain level of generality – as Mintz has described it, a typical meal composition of starchy 'core', a 'fringe' of pungent flavoured relish and legume is found across the globe (1992) – there are further sub-variants as to what counts as a meal. Ideas of a main meal can look dramatically different, only partly reflecting the availability of different staples in various parts of the world. Against this background, and given the human capacity for attributing meaning, foods are used, in the social anthropologists' expression, as a social marker. This mode of using food to register and label social boundaries applies both within and between social groups, where intra-group differences of social status, age, gender and more besides, are reflected in and reinforced by conventions identifying dishes or foodways deemed appropriate. Just possibly, inter-group differences may be thrown into sharper focus. Growing up in one or other social group entails learning typical definitions of food and typical foodways, allowing group members to take their own diet and eating arrangements for granted as 'right', 'only' or 'proper'. Contact with another group, however, may well sensitise people to what previously went unnoticed, sharpening alertness to difference.

Many 'Westerners' are familiar with the manner in which eating habits figure in the names one group uses to deride members of another: the French refer to the English as 'rosbif', the English call Germans 'Krauts' and North Americans talk of Italians as 'macaronis'. But the habit is not peculiar to the 'West' – the very word Eskimo meaning 'raw meat eater' derives from neighbouring Indians sneering at Inuit foodways (Farb and Armelagos 1980, p. 97) while at some stage, Uzbecks in Afghanistan have been called 'noodle eaters' by their neighbours and Arabs from Khuzestan known as 'lizard eaters' in Iran (Bromberger 1994, p. 185). Uncomplimentary as such name-calling may be, it serves as a way of placing a stranger, in short, an example of according them an identity – in this case one based on inter-group difference.

Inter-group differences themselves, together with the names that have been devised to describe them, have a history. For centuries of written history, social groups have encountered one another, a process steadily increasingly over the last few, with the rate accelerating in self-lubricating fashion to the point where globalisation (for the origins of the term see Beckford 2001) appears now to be discussed as if it were a peculiarly late twentieth century phenomenon. In parallel fashion, it seems to be easy to forget that the idea of a nation, together with nationalism, is comparatively recent – dateable from the end of the eighteenth century (Anderson 1991). All the same, historically earlier markers of boundaries between social groups such as region or religion and also ethnicity, have persisted across and within the nation-state (cf Zubaida 1994). Although ethnicity and nationality are self-evidently distinct, their relation and overlap need to investigated, and the mode in which each is analytically usefully to be defined (Bradby 1995; Banks 1996) the manner in which they ought to be distinguished is, for practical purposes, set aside in the present discussion. One key point must to be borne in mind, however. When contemplating their use as concepts for social analysis of food use as much as anything else, ethnicity and nationality, along with corresponding attributions of identity, are to be thought of as a matter of social interaction, a product of social relationships, arising when members of different groups encounter one another, even if they come to be deployed *as if* they were badges pinned to someone's lapel.

An association between ethnic/national identity and food is capturing increasing academic attention (e.g. Harbottle 1997; James 1997; Murcott 1996). More directly, it has been noted that the consequences of recognising differences in food habits across boundaries may be both positive – underscoring a sense of belonging to a group – as well as negative – vehicles for prejudice, xenophobia or chauvinism (de Garine 2001). Above all, perhaps, the best of the literature on food and national or ethnic identity (for a particularly fine contribution see the volume edited by Zubaida and Tapper 1994) illustrates Clifford's insistence referred to above that cultures do not stand still for their portraits to be drawn. Being able to spot shifts and continuing change, however, is helped by distinguishing between analysts seeking to provide an academic account and contemporary characterisations of ethnic or national identity in food and eating by those busy doing the eating, earning a living by buying and selling foods or even endeavouring to improve the population's health. Starting to separate academics' analyses from those of the social groups they study can be illustrated in an innovative approach to culture, nationality and food in one of the very few recent studies based in rural West Wales (Caplan et al. 1998). In this part of the Principality, ethnicity is intricately involved with the Welsh language. Residents in the study area could be broadly divided into non-Welsh speakers who define themselves as Welsh, English incomers including those who have learned and become fluent in Welsh, and those describing themselves as *Cymry Cymraeg* – Welsh and Welsh-speaking born and bred. Stressing that the relation between cuisine and ethnicity is also complex, Caplan argues that food *in* culture is to be separated from food *as* culture.

Food *in* culture describes a specifically Welsh mode of eating the researchers found among Cymry Cymraeg residents. Although it bore similarities to eating patterns elsewhere in Britain, dishes included those which are distinctively Welsh with dishes such as *cawl* (a meat and vegetable stew) or griddle-cooked Welsh cakes and a similarly distinctive pattern of food events through the day/week. These residents valued this style of eating with its plainness considered desirable. Correspondingly there was a resistance to what was thought of as 'foreign' foods, imports not just to Britain, but also to the locality. Food *as* culture stands in strong contrast to the quiet, private interiors of Cymry Cymraeg farmhouses. Food *as* culture refers to an image of Welsh cuisine created for public presentation, an invention for tourists, on restaurant menus, railway brochures and cookbooks. This is the Welsh food of picture postcards, of café advertisements to summer visitors for 'Welsh cream teas' with the addition of *bara brith* (a local currant bread) artfully to create a sense of national authenticity. While these examples are firmly commercial, the researchers found that Welsh residents also served food in this fashion albeit at home, but on those occasions made public when entertaining non-Welsh visitors.

Here, then are two rather different ways in which residents themselves both used food and gave foods a national adjective. Parallel variations are reported by members of the same team who undertook work on food use in what by anyone's definition is a multi-ethnic area of South London. Here residents descriptions of their own ethnicity primarily divide into White British or African-Caribbean/Black British (Caplan et al. 1998). A corresponding multiplicity of cuisines is evident, whether in terms of the range of local food shops and markets or in the dishes interviewees reported cooking/eating, but what counted as British or English food varied. Older residents regarded 'meat and two veg' or shepherd's pie English, favouring such dishes as plain and wholesome, whereas younger residents considered them bland and boring. By contrast, the 'culinary repertoires' of younger residents' belonging to both ethnic groups commonly include dishes of Italian, Chinese and Indian origin, dishes, moreover, thought of as unexceptionable and unlikely to be described as 'foreign'. There did seem, however, to be a disjunction between dishes attributed to these national origins, and West Indian foods.

Among African-Caribbean residents, an age difference could once again be found. Older people, especially those born in the Caribbean, inclined fairly consistently to hold on to culinary styles of their birthplace. Younger African-Caribbeans, however, reported eating more diversely across the week. At weekends, seeking to preserve a sense of original ethnic/national identity was evident. Sunday meals visiting parents would involve what one young man described as 'traditional West Indian food: rice and peas, carrot juice, pineapple juice' (Caplan et al. 1998, p. 180). Another talked of feeling 'more comfortable' by continuing to eat such food, while one woman is recorded as explaining:

'I try to keep a lot of my Jamaican in me. I make myself cook a West Indian meal on Sunday because I like to keep in touch or keep a part of my culture'.
(Caplan et al. 1998, p. 180).

This last serves as a reminder, if one be needed, that 'culture' is itself not a word which analysts adopt or possess for their exclusive use – any more than is the word 'ethnic', especially when applied adjectivally to foods, dishes or restaurants (cf Warde and Martens 2000, p. 81).

Sorting out academics' analytic usages from those of the social groups they study allows readier grasp of several features of culture, food and national identity. As suggested above, notions of national culinary traditions may be appropriated or annexed by tourist and 'heritage' industrial interests as part of their sales efforts – although it has been argued that there are non-material as well as economic rewards (Ten Eyck 2001). On the one hand, an appreciation of the historians' analyses allows that appropriation to be recognised for what it is. On the other, it paves the way for recognising that whether or not the origins are commercial, the characterisations of the labelling used by producers, manufacturers and distributors in the food chain tend to become adopted as established terminology and definitions. It is a common observation that several dishes with whose names a great many Britons have become familiar as Indian, and to the taste of which they have become accustomed, were invented in Britain to appeal to what was believed to be a British palate – dishes which have no counterpart anywhere on the Indian sub-continent. All the same, patrons of restaurants presented by their proprietors as Indian (often as not staffed by migrants from rural Bangladesh) will self-consciously consider themselves enjoying an exotic, non-English, ethnic meal. This is a practice which Lucy Long, a folklorist based in the US, has adroitly dubbed 'culinary tourism' (1998). By its nature, however, culinary tourism is liable to undermine itself. With repetition, the exotic supposedly sought by the tourist is converted to the mundane, the foreign is domesticated (Murcott forthcoming) – and the young South Londoners of Caplan and her colleagues' study, define pasta, pizza and chicken tikka as British foods.

The commercial appropriation of national adjectives for certain dishes and foods serves industrial interests of suppliers is one thing, the deliberate creation of cuisine as part of the creation of a nation arguably another. Although the two may shade into one another in a world market in which nations' economic interests in their balance of trade represent an integral part of nationhood, national identity and nationalism can in principle at least be distinguished from the commercial interests of one or other industrial enterprise located either within or across national borders. There is now a growing literature, much of it North American, documenting the manner in which the history of national identities can be investigated via a study of foodways (e.g. Pilcher 1998). National cuisines continue to be invented against but also associated with globalising trends (e.g. Wilk 2002) or dishes get transported by migrants to acquire their reputation as ethnic *thereafter* – for example, Gabbacia shows that bagels 'became firmly identified as "Jewish" only as Jewish bakers began selling them to their multi-ethnic urban neighbors' (Gabaccia 1998, p. 5).

If nothing else, contemplating the literature on food, nationality, ethnicity and identity leads to one inescapable conclusion: there can be no such thing as an authentic national cuisine. As Zubaida tartly observes, if an idea of 'the' standard Mediterranean diet exists at all, it does so as a very recent creation that is not found in any uniform fashion in that region of Europe. It is, rather, 'a modern construction of food writers and publicists in Western Europe and North America earnestly preaching what is now thought to be a healthy diet... (with) colleagues in Mediterranean countries... only too willing to perpetuate this myth' (Zubaida 1994, p. 43).

All the same, national and ethnic labels are pressed into service in order to convey a message in a fashion that may be equally highly compressed. It has, for instance been deployed as a means of managing and communicating food safety – specifically bovine spongiform encephalopathy (BSE) – in the marketplace. Despite presenting the public with dilemmas about choice when food shopping, people typically resort to several modes of reducing risk, one of which is to seek some proxy or indicator as a quick trustworthy basis for decision-making. Well known, this is one suppliers duly present: a label (Kjaernes 1999). That the Swedish knew to seek out beef labelled as home produced is not especially surprising (Murcott and Jansson 1999). Their herds had been reported to be BSE free, and opinion polling had reported a comparatively high level of trust in the state, its regulations and the quality of their enforcement.

But Macintyre et al. (1998) were surprised to discover a relaxed view of BSE among Scots – for the simple reason that BSE was considered an English not a Scottish problem. Once again, assuring safety was resolved when shopping by simply looking for the Scotch beef label. Each of these swiftly presented examples in their somewhat different ways, illustrates the manner in which members of the public find means of listening to expert advice, interpreting it for their local purposes, devising their own practical solutions, which cater for any residual doubts or mistrust they may have, whilst continuing the necessary tasks of provisioning for their households. From this somewhat different, practical perspective, suppliers have successfully used a national identity as a shorthand applied to their food products to support sales. As Mary Douglas might have put it, here indeed is a condensed symbol.

2.9
Food, Social Hierarchy and Class Cultures

The social and historical variability as to what counts as food can also be discerned when turning to consider food and social hierarchy. Historical evidence suggests that in fourteenth and fifteenth century France, local peasant cuisines were both distinct from each other as well as from the more uniform cuisine of the elite: correspondingly 'escargots, the delicacy of the higher estates were considered inedible by the lower ones' (Falk 1994, p. 91). A close parallel can be found for 1930s England, in which the shoots of swede (rutabaga in the US and turnip in Scotland) cooked as a vegetable were, according to an Oxfordshire shepherd, considered a 'great delicacy' by the gentry, whereas '"Us poor folk don't trouble about 'em"' (White 1932, p. 217). The patterns, though, are not simple and such distinctions do not remain static. The Harvard anthropologist Theodore Bestor, for instance, shows how judgements of raw fish in the form of sushi, have shifted from 'an exotic, almost unpalatable ethnic specialty, then to haute cuisine of the most rarefied sort... (to) become not just cool, but popular' (Bestor 2000).

It must not, however, be assumed that an haute cuisine can universally be distinguished from more plebeian or peasant cuisines. Such a contrast is not evident in most African 'traditional', pre-colonial societies, which, as Jack Goody has shown, is not simply a matter of degrees of social or political complexity. Based on a comparison between two North Ghanaian tribes among whom he has conducted extensive fieldwork, he demonstrates that there are significant differences between the two tribes. The Lo Dagaa have no tradition of chiefship, whereas the Gonja have a long history as an independent kingdom with a ruling estate, commoners, as well as slaves. The geographical environment and consequent agriculture of each tribe, differs markedly, as do each tribe's rules of hospitality, exchange and ceremony – in all of which food is implicated. The basic diet for each is also contrasted, even if conforming to Mintz' core-fringe-legume composition noted above. Among the Lo Dagaa, Goody finds little daily variation in the diet of a millet porridge, accompanied by a leaf and soup of ground-nuts. For the Gonja, the diet consists of yams or cassava and grains depending on the time of year, with some fish or wild meat. Yet – and this is the critical point - despite such differences, and despite marked differences in the detail foodstuffs composing each cuisine, the shape or form of those cuisines was strikingly similar, with 'little differentiation of cooking… in relation to the different strata of Gonja society' (Goody 1982, p. 96). Goody concludes, then, that the absence of a distinct high and low cuisine of a centuries-long kind familiarly found across Europe, is a matter of the *type* of status differentiation. The African cases are examples of what he calls hieratic systems, as distinct from those he describes as hierarchical. The former have minimal variation in way of life, attitudes and so on across the whole society, whereas the latter – e.g. Indian castes, Mediaeval feudal estates, as well as the class structures of modern industrialised societies – 'have well developed, stratified sub-cultures' (Goody 1982, p. vii).

Identifiable as part and parcel of such sub-cultures are Europe's intricate 'culinary cultures' (the shorthand expression introduced earlier in this chapter), the collections of attitudes and tastes associated with food, cooking and eating. The slowly emerging sociological-historical study of these facets of food and culture is beginning to document different degrees of intricacy of the last four centuries in culinary cultures and sub-cultures between and within what eventually emerged as modern European nation-states. Although by comparison with the African cases Goody reports, the culinary cultures of France and England for instance, can seem very similar, there are significant differences – captured in a persisting tendency to use the French expression haute cuisine rather than an English translation. For there is an intellectual puzzle: what led to the emergence of an elite or haute cuisine in France but no obvious counterpart in England? Tackling it is complicated by the historical fact of overlaps between cookery in France and England, with the English upper classes' adoption of the former at several stages over at least the last century.

The answers proposed are similarly complicated, requiring a so-called developmental perspective – one which entails a dissatisfaction with the emphases of Lévi-Strauss or Mary Douglas, and, in particular, their neglect of history and *social* development. Stephen Mennell has promoted this perspective when tackling that intellectual puzzle about French and English culinary cultures (1985). Attention, he argues, has to be focussed on the differences in the social processes and consequences of competition that are part and parcel of complex patterns of social stratification, describable systems of the hierarchical relation between the various social statuses of which the systems are composed.

It means tracing changes from the culinary cultures of Mediaeval Europe to take account of the consequences in England of the Civil Wars and in France the later Revolution. In the Middle Ages, distinctions between culinary cultures ran across European geographical boundaries with contrasts lying between the aristocracies of no matter which region on the one hand and the successively lower feudal orders on the other. The gradual formation of nation-states was paralleled not only by the emergence of what could more or less be contrasted as 'national' cuisines, but by differing degrees of differentiation *within* the borders of those nation-states. Mennell argues that it was the contrasting nature of the court in England and France, the connection of their respective aristocracies to their country estates together with the different balances of power and habits of deference between them and each monarchy, to which scholars must look to explain the emergence of the self-conscious French traditions of haute cuisine. Mennell's readable history points out, of course, the manner in which even the more muted haute cuisine of England, never mind that of France does not remain static. What counts as the latest in elite eating at different periods is subject to successive waves, especially in France, of increasing sophistication followed by a revolt which once again extols the virtues of simplicity: the *nouvelle cuisine* of the 1970s was not new but preceded by an identically named *nouvelle cuisine* of the previous century.

Complicated enough to document for two countries, never mind the further work of this kind still to be undertaken for others, Mennell, among others, argues that appetite and taste – both literally and metaphorically – are as much *socially* as biologically shaped. Mediaeval aristocrats' use of the sheer volume of food at their disposal as an aspect of their means for displaying social superiority to their inferiors gave way to an emphasis on greater discrimination. This was paralleled and made increasingly possible by and consequent upon greater social stability and security in food supplies across Europe, combined with extensions of trade and the globalisation of sixteenth century explorations. Once the idea of refinement and discrimination at the table became evident, continued improvement in the security of supply meant that wider sections of society moved into a position of being able to emulate this aspect of elite eating. The idea of eating vast quantities became vulgar while more elaborate foods, that could be qualitatively gauged became the height of good taste as the mode in which higher social groups asserted the distance between themselves and those beneath them.

Such a use of food (among other material goods) as a means of asserting social superiority in good taste is often enough remarked upon by academics (e.g. Falk 1994; Essemyr 1992; Atkins and Bowler 2001), the conspicuous consumption of the 'vicarious leisure' class which Thorstein Veblen detailed even within the servant classes (Veblen 1970 [1925], p. 60), never mind the well worn

English phrase 'keeping up with the Joneses'. Attempts have also been made to identify the social mechanisms whereby a psychology and associated behaviour of emulation are created. One such is the notion of competitive struggle, coined by the late Pierre Bourdieu in his major study of culture, status and distinction (1984, p. 165). Showing that taste in food (along with music and art) is not simply correlated with income differentials, but associated as closely if not more so with levels of education, he argued that investment in and accumulation of 'cultural capital' is key in the social creation of taste, reflected in whatever fashions in food as much as in décor or clothes dominate at the time. Correspondingly, those with limited cultural capital lower down the social scale are excluded, held to lack the capacity to discriminate, unable adequately to appreciate, even while wishing to emulate, the tastes displayed by those above them.

Such a major contribution as Bourdieu's inevitably involves a correspondingly major secondary literature (e.g. Jenkins 1992) and scepticism as to the extent to which his analyses apply outside his native France (e.g. Wood 1995). All the same, it is possible to find evidence which in certain respects has strong parallels with his findings. What could be considered a precursor is found in Littlejohn's (1963) undeservedly neglected single-handed study of a small village in rural Scotland. His central concern derived from sociological questions about the dynamics of socio-economic hierarchies and associated mode of living, styles and tastes. A segment of the study concerns meal times, particularly tea at which all members of the family in all social classes are usually present. The difference in behaviour patterns, culturally defined expectations and the disposition of material items, including foods, not only illustrate marked contrasts in taste between classes but also class related gender differences – a feature also emphasised in Bourdieu's analyses. Contrasts in the class cultures between the highest and lowest ranks of the village can be summarised as follows:

In the highest ranking class, everyone sits down at the table together, including the woman of the house. At her side are the teapot, milk jug, sugar bowl and all the cups. Everyone waits for her to invite them to begin eating, and it is very bad manners for a visitor to help themselves to any food on the table. Members of the family are also to wait for the woman to invite them to eat. If they wish for something they have to ask for it, adding 'please'. The food itself is so arranged as to facilitate the woman's presiding over the meal: bread is cut thinly, often made into minute sandwiches. Scones are small, and large cakes are put on the table uncut, so that if anyone wants a piece of cake he or she can only have it after the woman has specially cut one for them.

In the lowest ranking class tea starts with everyone except the woman sitting at the table. She walks round the table pouring tea into the empty cups she has earlier set beside each place. Only then does she sit down. The milk and sugar are in the centre of the table and each person simply reaches out his hand for them and adds whatever they wish to their cup. The food is in the centre of the table too and each person reaches out a hand, takes what they want without reference to the wife or anyone else. If some plate of food is beyond arm's reach, a person takes their knife and stabs the desired item on the serving plate and brings it onto their own plate. Again, the food is arranged specifically to fit into this pattern of activity: bread is cut thickly and is never made into sandwiches. Scones are large, and cakes are put on the table already cut into individual portions. Should anyone want more tea, they simply pass their cup to the woman of the house, without using any words…

Littlejohn's work not only illuminates what to many familiar with Britain are recognisable contrasts, it also demonstrates the way in which food, eating, provisioning, conversation, domestic divisions of labour and more are intimately related in the complexities of routine day to day activity.

Littlejohn's study is based on extended fieldwork undertaken in the years just after the Second World War. The particular details are liable to have altered in the half century and more since then, but there is strong evidence (e.g. Warde 1997, 2000) for the perpetuation of correlations between food and culture, taste in eating, and social class. By way of rounding off this brief overview of food, culture and social class, even sketchier mention is made of evidence going still further back in time. The purpose is speculatively to pave the way for discussions of the relation between higher/lower class cultures and the 'cultures' of expert science and scientifically little- or un-trained publics (see Collins forthcoming for the strong justification for use of this cumbersome expression rather than lay publics). How far is it possible to use the relation between higher class value judgements of lower class activities as a model for understanding the relation between expert science and publics without formal scientific qualifications? The question is only posed while turning to refer rapidly to the nineteenth and early twentieth century case of fish and chips – working class use and middle class condescension – of which Walton's book length treatment must be one of the most exhaustive (1992).

Suppliers, specifically the fish friers, struggled for respectability. For their products were identified with slum dwellers, reputedly dubious hygiene and denouncements of the supposedly intemperate domestic budgeting of incompetent housewives who resorted to fish and chips to feed their families. Middle class manners were offended by the public consumption of fish and chips, wrapped in newspaper and eaten with presumably unwashed fingers. While Walton finds evidence for such judgements of the trade and its commodity in the inter-war period, he traces evidence for such an unfavourable reputation to its earliest years.

Attacks on fish and chips from Medical Officers of Health, the popular press and assorted civic dignitaries and representatives of officialdom began in earnest at the turn of the century and persisted throughout the period and beyond it.

(Walton 1992, p. 148)

It is the alliance between medical professionals – and the associated appeal to nutritional science – opinion formers and others in positions of power which is of significance in the present context. There were strong medical opinions condemning fish and chips: sustained efforts, for instance, were made in the early part of the twentieth century to link fish and chips with typhoid fever, and later, to implicate the diet in high rates of infant mortality. Assessing his evidence (presented in far more detail than is possible to recapitulate here) Walton is led to draw on the comments of the historian Charles Webster that

> *when dealing with working-class health and living standards Medical Officers of Health were all too likely to parade the prejudices of their class as if they were scientific medical judgements.*
>
> (Walton 1992, p. 150)

Such an assessment did not, though, apply to all medical professional commentators, and others recommended fish and chips as of nutritional value as well as aesthetic respite from an otherwise monotonous diet of tea, bread and dripping. Medical commentators who arrived as this more favourable judgement may well have been those who took a closer, more dispassionate look at the culinary cultures of the class which their fellows too hastily condemned. In turn, so doing is commended to any of us concerned to understand what seem to be gaps between the advice of experts and those to whom it is addressed...

2.10 Concluding Remarks

A set of questions was posed at the beginning of this chapter. Is there a meaning conveyed in so much of recent talk of 'culture' and food that cannot be captured by the use of other expressions, or is food culture no more than a fashionable cliché – convenient, but cliché none the less? Accordingly one purpose of this chapter has been to provide a basis for considering whether paying a little attention to some of the social anthropological/sociological work might offer the prospect of a more extensive and nuanced appreciation of food and culture. It has sought to do so by no more ambitious a means than presenting a series of examples – with little to be gained in attempting a summary of them at this final stage. How far and in which directions they do indeed hold out the promise of a developed appreciation requires further discussion. At the least, the starting point for that discussion is likely to agree that the sections above illustrate the way that 'culture and food' has been approached in several very different ways. Arguably, each takes the discussion further than using culture as a synonym simply for lifestyle or maybe ethos. A related purpose is to pave the way for more fully grounded understandings of what is so commonly discussed as a mismatch between expert commentary and public attitudes and behaviour, in food as in much else – a point taken up elsewhere in this volume. While the utility of 'food and culture' awaits more thoroughgoing assessment, in the meantime the foregoing does, however, allow one or two last basic observations to be made about culture in general, and in relation to food in particular. First, everyone (with only the not quite mythical feral child as an exception) is socialised – i.e. brought up – in one or other cultural context. Culture is not simply some sort of property peculiar only to strangers or to unfamiliar ways of life, and – along with ethnic – is not helpfully used as an adjective meaning exotic. Second, to record that food use varies statistically in terms of culture where culture is operationalised along with age, or sex, ethnicity or area of residence – as if each were a badge – is important in providing preliminary signposts to patterns of consumption or whatever is being studied. Equally, though, it is important to recognise that regarding culture as if it were an attribute of an individual or to list it as one of a number of social variables, is not the conclusion of, but the starting point for analytic investigation. Finally, it is probably high time that culture is no longer analysed as static and inflexible, simply described as transmitted immutable from generation to generation, but as complex and dynamic, to be understood historically, subject to and product of human creativity.

Acknowledgements

For reasons outwith my control, work on this chapter has been fitful and unduly protracted. I am most grateful to the editors for their patience and thoughtful support. I should also like to thank: Sue, Carolyn, Ricky, Nicola, Kath, Terry and also David for helping me through a tricky time; Warren and Amy Belasco in Washington DC for their good conversations and extended hospitality on September 11, 2001 while working on this chapter; Rachel Burr and Matt Freidson in London for listening and making especially helpful suggestions; and John Murray-Browne, proprietor of the Angel Bookshop in north London, for his enviable knack of asking the apt question.

References

Secondary Sources

Anderson B (1991) Imagined Communities: reflections on the origin and spread of nationalism. Verso, London

Atkins P, Bowler I (2001) Food in society: economy, culture, geography. Arnold, London

Atkinson P (1978) From honey to vinegar: Lévi-Strauss in Vermont. In: Morley P, Wallis R (eds) Culture and curing. Peter Owen, London

Atkinson P (1983) Eating virtue. In: Murcott A (ed) The sociology of food and eating. Gower, Aldershot

Banks M (1996) Ethnicity: anthropological constructions. Routledge, London

Beckford JA (2001) Developments in the sociology of religion. In: Burgess RG, Murcott A (eds) Developments in sociology. Pearson, London

Bestor TC (1999) Wholesale sushi: culture and commodity in Tokyo's Tsukiji Market. In: Low S (ed) Theorizing in the city. Rutgers University Press, New Jersey, US

Bestor T (2000) Foreign Policy, November-December pp 54–63

Booth DA (1994) Psychology of nutrition. Taylor Francis, London

Bourdieu P (1984) Distinction a social critique of the judgement of taste. Routledge and Kegan Paul, London

Bradby H (1995) Ethnicity: not a black and white issue. A research note. Sociology of Health & Illness 17: 405–17

Bromberger C (1994) Eating habits and cultural boundaries in Northern Iran. In: Zuabaida S, Tapper R (eds) Culinary cultures of the Middle East. I. B. Tauris, London

Brooker P (1999) A Concise glossary of cultural theory. Arnold, London

Bush H, Williams R, Bradby H, Anderson A and Lean M (1998) Family hospitality and ethnic tradition among South Asian, Italian and general population women in the West of Scotland. Sociology of Health & Illness 20: 351–80

Caplan P, Keane A, Willetts A, Williams J (1998) Studying food choice in its social and cultural contexts: approaches from a social anthropological perspective. In: Murcott A (ed) The nation's diet: the social science of food choice. Longman, London

Charles N, Kerr M (1988) Women, food and families. Manchester University Press, Manchester

Cherikoff V, Brand JC (1988) Is there a trend towards indigenous foods in Australia? In: Truswell SA, Wahlqvist Mark L (eds) Food habits in Australia. René Gordon, N. Balwyn, Victoria

Clifford J (1986) Introduction: Partial Truths. In: Clifford J and Marcus GE (eds) Writing culture: the poetics and politics of ethnography. University of California Press, Berkley, CA

Counihan C, van Esterik P (1997) Food and culture: a reader. Routledge, New York

Crotty PA (1988) The disabled in institutions: transforming functional into domestic modes of food provision. In: Truswell SA, Wahlqvist ML (eds) Food habits in Australia, René Gordon, N. Balwyn, Victoria

De Garine I (2001) Views about food prejudice and stereotypes. Social Science Information 40

Douglas M (1966) Purity and danger. Routledge and Kegan Paul, London

Douglas M (1972) Deciphering a meal. Daedalus 101: 61–81

Douglas M (1975) Social and religious symbolism of the Lele. In: Douglas M (ed) Implicit meanings. Routledge and Kegan Paul, London

Douglas M (1977) "Beans" means "thinks". Listener 8: 292–293

Douglas M (ed) (1984) Food in the social order. Russell Sage Foundation, New York

Douglas M, Nicod M (1974) Taking the biscuit: the structure of British meals. New Society 30: 744–747

Ekström MP, Fürst EL (2001) The Gendered Division of Cooking. In: Kjaernes U (ed) Eating Patterns: a day in the lives of Nordic peoples. SIFO (National Institute for Consumer Research) Lysaker, Norway

Essemyr M (1992) Nutritional needs and social esteem: two aspect of diet in Sweden during the 18th and 19th centuries. In: Teuteberg H (ed) European food history: a research review. Leicester University Press, Leicester
Falk P (1994) The consuming body. Sage, London
Farb P, Armelagos G (1980) Consuming passions: the anthropology of eating. Houghton Mifflin, Boston
Fieldhouse P (1986) Food and nutrition: customs and culture. Chapman & Hall, London
Gabaccia DR (1998) We are what we eat: ethnic food and the making of Americans. Harvard University Press, Cambridge, Mass.
Geertz C (1973) The interpretation of cultures. Fontana, London
Gofton L, Murcott A (2001) The special place of chocolate in the Anglo-American diet: towards a sociology of food cravings and addictions. In: Hetherington, M (ed) Food cravings & addiction. Leatherhead International, Leatherhead, UK
Goody J (1982) Cooking, cuisine and class. Cambridge University Press, Cambridge
Harbottle L (1997) Fast food/spoiled identity: Iranian migrants in the British catering trade. In: Caplan, P (ed) Food, health and identity. Routledge, London
James A (1981) Confections, concoctions and conceptions. In Wates B et al. (eds) Popular culture: past and present. Croom Helm, London
James A (1997) How British is British food? In: Caplan P (ed) Food, health and identity. Routledge, London
Jenkins R (1992) Pierre Bourdieu. Routledge, London
Keane A, Willetts A (1995) Concepts of healthy eating: an anthropological investigation in south-east London. Goldsmiths' College Working paper, University of London, London
Kemmer D, Anderson AS, Marshall DW (1998) Living together and eating together. The Sociological Review 46: 48–72
Kahn JS (1995) Culture, multiculture, postculture. Sage, London
Kjaernes U (1999) Food risks and trust relations. Sociologisk Tidsskrift 7: 265–284
Kyle R (2001) Children, food and health: a report on the first six months work in Sandwell. Sandwell Health Partnership, Birmingham
Leach E (1970) Lévi-Strauss. Collins, Glasgow
Leach E (1976) Culture and communication. Cambridge University Press, Cambridge
Lévi-Strauss C (1964 [1970]) The raw and the cooked. Cape, London
Lévi-Strauss C (1966 [1965]) The culinary triangle. New Society: December 937–40
Littlejohn J (1963) Westrigg: the sociology of a Cheviot village. Routledge and Kegan Paul, London
Long LM (1998) Culinary tourism: a folkloristic perspective on eating and otherness. Southern Folklore 55: 181–204
Lupton D (1996) Food, the body and the self. Sage, London
Macintyre S, Reilly, J, Miller D, Eldridge J (1998) Food choice, food scares, and health: the role of the media. In: Murcott A (ed) The Nation's Diet: the social science of food choice. Addison Wesley Longman, London
Mennell S (1985) All manners of food. Blackwell, Oxford
Mennell S, Murcott A, van Otterloo A (1992) The sociology of food: diet, eating and culture. Sage, London

Mintz SW (1985) Sweetness and power: the place of sugar in modern history. Viking, New York

Mintz SW (1992) Die Zusammensetzung der Speise in frühen Agrargesellschaften Versuch einer Konzeptualisierung. In: Schaffner M (ed) Brot, Brei und was dazugehört Chronos, Zürich

Murcott A (1982) On the social significance of the "cooked dinner" in South Wales. *Social Science Information* (Anthropology of Food Section) 21: 677–695

Murcott A (1983a) "It's a pleasure to cook for him…": food, mealtimes and gender in some South Wales households. In: Gamarnikow E, Morgan D, Purvis J, Taylorson D (eds) The Public and the private. Heinemann, London

Murcott A (1983b) Cooking and the cooked. In: Murcott A (ed) The sociology of food and eating. Gower, London

Murcott A (1996) "Food as an expression of national identity". In: Gustavsson S, Lewin L (eds) The future of the nation state: essays on cultural pluralism and political integration. Nerenius – Santérus, Stockholm, Sweden

Murcott A (forthcoming) Modes of eating the other: on the analytic utility of 'culinary tourism'. In: Long, L (ed) Culinary tourism: eating and otherness. University of Kentucky Press, Kentucky

Murcott A, Jansson S (1999) Theorising ethnicity and nationality in food and eating: examples from Britain and Sweden' Paper presented at 'Crossing Borders: Food and Agriculture in the Americas' Joint 1999 Annual Meetings of ASFS & AFHVS and the annual Food Choice conference, Ryerson Polytechnic University,Toronto, Ontario, Canada June 3–6

Pilcher JM (1998)"Que vivan los tamales": food and the making of Mexican identity. University of New Mexico Press, Albuquerque

Pill R, Stott NCH (1982) Concepts of illness causation and responsibility. Social Science & Medicine 16: 43–52

Pill R, Stott NCH (1985) Preventive procedures and practices among working class women: new data and fresh insights Social Science & Medicine 21: 975–983

Pill R (1983) An apple a day… some reflections on working class mothers' views on food and health. In: Murcott A (ed) The sociology of food and eating. Gower, Aldershot

Sills DL (ed) (1968) International encyclopaedia of the social sciences. Macmillan & Free Press, New York

Symons M (1982) One continuous picnic. Penguin, Ringwood, Victoria

Ten Eyck TA (2001) Managing food: Cajun cuisine in economic and cultural terms. Rural Sociology 66 : 227-243

Tyler EB ([1871] 1958) Primitive culture. Smith, Gloucester, Mass

Veblen T (1970) The theory of the leisure class. Unwin, London

Walton JK (1992) Fish and Chips and the British working class 1870–1940. Leicester University Press, Leicester

Warde A (1997) Consumption, food and taste: culinary antimonies and commodity culture. Sage, London

Warde A, Martens L (2000) Eating Out: social differentiation, consumption and pleasure. Cambridge University Press, Cambridge

Williams J (1997) "We never eat like this at home": food on holiday. In: Caplan P (ed) Food, health and identity. Routledge, London

Williams R (1983) Keywords: a vocabulary of culture and society. Fontana, London (revised edition)

Wilk RR (2002) Food and nationalism: the origin of "Belizean food" in Belasco. In: Warren, Scranton P (eds) Food nations: selling taste in consumer societies. Routledge, New York

Wilson CA (1973) Food and drink in Britain. Penguin Books, Harmondsworth

Wood RC (1995) The Sociology of the meal. Edinburgh University Press, Edinburgh

Yudkin J, McKenzie JC (eds) (1964) Changing food habits. McGibbon and Kee, London

Zubaida S (1994) National, communal and global dimensions in middle eastern food cultures. In Zuabaida S, Tapper Ri (eds) Culinary cultures of the Middle East. I. B. Tauris, London

Zuabaida S, Tapper R (eds) (1994) Culinary cultures of the Middle East. I. B. Tauris, London

Primary Sources

The Independent on Sunday (2001) Magazine section, 8 July: 43 (Michael Bateman 'If you can't stand the heat…')

Cooper D (2001) 'The Food Programme' BBC Radio 4, 12 August 2001

White F (ed) (1932) Good things in England. Jonathan Cape, London

Science, Society and Public Confidence in Food Risk Management

3

LYNN FREWER

3.1 Introduction

Many members of the public have become increasingly concerned about risk management practices, particularly in the context of food and food production. This trend has manifested itself through expressions of concern and anxiety about changes in agricultural practices and food production technologies. Perceptions of risk associated with genetically modified foods, Bovine Spongiform Encephalopathy (BSE) and Creutzfeldt-Jakob Disease (vCJD), emerging pathogens such as E-Coli 0157, and increasingly complex information about appropriate nutrition have all been, and continue to be, foci of public fear and cynicism about how food risks are managed.

In the past, risk assessment and risk management have tended to create firewalls between natural and social science input into risk analysis. However, increased understanding of the social and cultural factors that influence people's responses to different food hazards will provide the most rigorous basis from which to align risk management with the needs of the wider public. It is also important to facilitate the process of integrating natural science with society in a way that promotes quality of life and environmental sustainability, and this can only occur if social as well as technical issues are included in debate about risk issues. In the area of risk management, this process of integration will only occur if understanding of the cultural systems that define people's responses and representations of risk provide input into the debate about how to manage risk and risk mitigation priorities. Failure to develop such interdisciplinary integration is likely to increase public distrust in science and its associated technological applications, and reduce public acceptance of risk management decisions. It is important to ask why members of the natural science community have historically argued that research within the natural science area is immune from influence from social processes, particularly in the area of risk management and selection of priorities for research strategies (Woolgar 1996). If questions such as these are not asked, it is likely that public cynicism regarding the motives of scientists, risk regulators and risk managers will continue.

3.2
An Historic Perspective on Risk Perception

In the 1970s, regulatory agencies and other institutional actors with responsibility for the management of risks believed that the general public were acting "irrationally" with respect to their responses to different technological and lifestyle hazards. For example, people's rejection of nuclear technology was assumed by technical risk experts in this area to be "irrational" as probabilities of personal harm were relatively low compared to other hazards linked to lifestyle choices to which people exposed themselves on a voluntary basis. For example, it was argued that an individual was more likely to experience the negative consequences of the risks associated with a car accident than from an incident involving a nuclear installation. Despite this difference in probabilities, people expressed much greater concern about the risks of the latter relative to the former. Technical risk assessors argued that if only people could understand the science that informed technical risk assessments in the same way as did experts, then "irrational" responses and reactions would disappear. For example, the general public would not be concerned about the development and application of an emerging technology, or be unduly concerned about risks with relatively low occurrence. The application of arguments of this sort has permitted elite groups to dismiss such public reactions as inappropriate and irrelevant at best, and symptomatic of public "ignorance" and Luddite responses to technology policy at worst. The membership of decision-making bodies was consequently restricted to elite groups of "right-thinking" individuals who had the skills and intellectual capability to respond in an "appropriate" manner to technical risk information provided by probabilistic risk assessment. This attitude has had, and to some extent continues to have the effect of promoting scientific agendas in a way that is independent of the concerns or priorities of the rest of society; however it is these public concerns that have influence on, and consequences for, human health, food availability and sustainability, economic growth and international regulation. Understanding of public concerns and priorities, and associated dynamic shifts with time change should form the basis for the development of an effective risk management strategy. Moreover, effective risk communication in itself is unlikely to be enough to allay public concerns about risk and risk management. Greater emphasis should be placed on actively involving the public in the process of decision-making regarding risk management practices, ensuring that these processes are an explicit, rather than implicit, part of the culture of institutions involved in regulation.

3.3 Risk Perception and Communication

Initial research into risk perception was conducted in the late 1970s and 1980s. The seminal work by Paul Slovic and colleagues (e.g. Slovic 1993) resulted in the development of a theory of risk perception known as the "Psychometric Paradigm". This research posited that people judge "risk" in terms of psychological dimensions other than probability and harm: in particular, perceived control, the extent to which a risk was perceived to be taken on a voluntary basis, dread, and catastrophic potential were found to be important psychological determinants of people's responses to risk. These psychological factors were used to explain public responses to low probability technological risks. For example, it was argued that people objected to nuclear power because they had no personal choice over whether or not to expose themselves to the risks, and believed that if a nuclear accident occurred the risks were potentially catastrophic. The existence of alternative energy sources also resulted in people questioning whether nuclear energy was, in fact, necessary. In comparison, driving a car was a voluntary and controlled activity that provided a direct benefit to the person engaging in the potentially risky activity.

Research within the "psychometric paradigm" provided the foundation for empirical investigation into qualitative differences between "lay" and "expert" conceptualisations of risk. It has been argued that many problems associated with effective risk management relate to differences in the way experts and lay people conceptualise risk (e.g. Barke and Jenkins-Smith 1993; Flynn, Slovic, and Mertz 1993; Lazo, Kinnell, and Fisher 2000; Mertz, Slovic, and Purchase 1998). Specifically, differences in concern between experts and the general public have been explained by the hypothesis that expert risk perception is not affected by the psychological factors that appear to drive the concerns of non-experts. That is, experts are less likely to consider that factors such as voluntary exposure, dread, or potentially catastrophic consequences have legitimate input into risk management policies. However, some authors have proposed that these reported differences between lay and expert groups are the result of methodological weaknesses that produce systematic bias in the results of comparisons of perceptions of risk derived from expert and non-expert groups. For example, Rowe and Wright (2001) have published a critique of empirical evidence supporting the psychometric paradigm on the basis of methodological artefact, arguing that observed differences are really the result of demographic differences between expert and non-expert groups. As a case in point, a relatively larger percentage of women may be allocated to the "non-expert" group, and women are known to rate risks as being more serious than do men (Kraus et al. 1992). Thus it is easy to dismiss the views of women as "non-normative" and maintain the views of the dominant expert elite as the foundation for risk management practices.

An alternative view is that, despite the demographic homogeneity of expert groups relative to the general population, this homogeneity realistically reflects an important social reality. If risk management and risk communication are driven by expert groups, and the composition of these groups favours a dominant demographic segment of the population, then there will be real consequences for society in terms of how communication processes are operationalized, and the infra-structure associated with how risk management is organised. The gap between science and society will continue to widen, and the public continue to become more distrustful of elite groups which promote an understanding of risk very different from the majority of the population.

Alternative arguments have focused on differences in the extent to which experts and the general public estimate the magnitude of a risk associated with a particular hazard. For example, Sjöberg et al. (2000) report that experts have a similar attitudinal structure to the public, but differ drastically in level of perceived risk. However, the authors note that differences may be hazard specific, (in this case, the hazard under investigation was nuclear energy). This finding cannot be generalised to all hazards, particularly those where the "technical expert" does not possess technical expertise specifically relevant to the hazard being assessed. Finally, it is arguable that risk-related concerns (whether held by lay people or experts in a specific area of risk assessment) may be unique to a particular hazard domain (Frewer et al. 1996). Differences in risk perception between expert and lay communities may differ qualitatively rather than quantitatively, (Larkin 1983; Hunt and Frewer 1998; Miles and Frewer 2001).

Use of qualitative analysis has supported this hypothesis in the case of organophosphate sheep dip (Carmody et al., submitted). The perceptions and beliefs held by three stakeholder groups, farmers, technical risk experts drawn from the biosciences, and the general public, were analysed using semi-structured interview techniques. The results indicated that, while all three interest groups shared the most often expressed concerns, the way and frequency with which these were associated varied between the three groups. In particular, values (for example, environmental concern, anti-technological perceptions) were most important for the general public relative to the other two groups. Thus the research provided evidence to support the idea that experts and lay people think differently about a given risk, but also provided evidence that those who experience direct benefits from a technology hold qualitatively different representations of beliefs and concerns compared to those who do not stand to benefit from its development and application. Of course, the results cannot be generalised to all hazard domains without further research. In particular, differences in public perceptions of different food hazards across nutritional and technological domains (Sparks and Shepherd 1994) means that generalization of results may be difficult and that domain specific analysis should be carried out as standard practice.

There is also evidence that, particularly but not only in the area of technology and its applications, people will tolerate risk if they perceive there is some direct benefit to themselves (Alkhami and Slovic 1994; Frewer et al. 1998). People are more concerned about the extent of the personal or environmental benefits resulting from a particular technology than the extent of the risks. Public acceptance will be driven by perceptions of personal benefit (Frewer et al. 1997; Frewer 2000; Deliza 1999; Da Costa et al. 2001). In terms of acceptance of food risks, potential benefits of consuming different products are likely to vary. For example, sensory properties, local traditions in cuisine, and cultural beliefs such as appropriateness of use of a particular food, may represent more important benefits to consumers in comparison to producer benefits or even improved nutrition.

The relationship between perceived risk and benefit is not necessarily straightforward. Consider the case of genetically modified foods. Siegrist (1999) reported that the extent to which people trust companies and scientists performing gene manipulations influences perceptions of risk and benefit associated with the use of the technology in food production. When trust is controlled statistically in the analysis of the data, the inverse relationship between perceived risk and benefit vanishes. Furthermore, perceived risk and benefit were reported to contribute independently to technology acceptance. The focus of this study was therefore geared towards understanding how well people thought the risks of genetic modification were managed ("societal trust") rather than focusing on how truthful institutions were about the extent and impact of the risks themselves.

Results such as these have lead to the formation of a hypothesis that trust in institutions is a causative factor linking technology acceptance and public confidence in science and regulatory practices (Czetovitch and Lofstedt 1999). It is, however, necessary to distinguish between trust in institutions (societal trust) and trust in information and information sources (source credibility). Societal trust may have a different relationship with perceptions of risk and benefit compared to people's beliefs in the honesty of the same institutions as sources of information. Empirical work in this area has explicitly attributed information to a particular source (often as an experimental manipulation), and gauged the effect on perceived risk and benefit. Trust is usually assumed to be multidimensional and specifically influential with respect to different sources and subjects of communication (Frewer et al. 1997; Johnson 1999). The impact of the information (or informational content) of a risk message on trust in an information source can also be measured post information intervention. Hovland, Janis and Kelley (1953) have identified two important dimensions that contribute to the extent to which people trust information sources; expertise and trustworthiness. Expertise refers to the extent to which a speaker is perceived to be capable of making correct assertions, whilst trustworthiness refers to the degree to which an audience perceives the assertions made by a communicator to be ones that the speaker considers valid. An example is provided by the case of genetically modified foods (Frewer et al., submitted). Two kinds of information about genetically modified food were presented to participants in an intervention trial – "product specific" information (which was skewed to present genetically modified foods in a positive light) and "balanced" information, (which discussed the potential risks and benefits of genetic modification of foods in a very neutral, but probabilistic, way). The information was attributed either to a consumer organization (shown to be highly trusted in pilot research), or an industry association (shown to be highly distrusted), or the European Commission (moderately trusted) under the different experimental conditions used in the study. Attitudes towards genetically modified foods were assessed before and after the information intervention. Data about people's perceptions of information source characteristics were also collected. The results indicated that the extent to which people trusted information sources had little impact on attitudes towards genetically modified products or product acceptance. Prior attitudes towards genetically modified foods accounted for almost 95 and 90 per cent of the variance in perceived benefit and perceived risk respectively – trust, however, had negligible impact on these risk related attitudes. The extent to which participants trusted the information sources was predominantly determined by participants' already existing attitudes to genetically modified foods, and not influenced by perceptions of source characteristics. In other words, independent of the type of information provided, information provision in itself had little effect on people's attitudes towards genetically modified foods. Furthermore, perceptions of information source characteristics did not contribute to attitude change, nor did the type of information strategy adopted have an impact on post-intervention attitudes. Of greatest concern to industry and other institutions with an interest in information dissemination was the observation that the extent to which people trusted the information sources appeared to be driven by people's attitudes to genetically modified foods, rather than trust influencing the way that people reacted to the information. In other words, attitudes were used to define people's perceptions regarding the motivation of the source providing the information. This perhaps is understandable in the case of the product specific information, which was very positive about genetic modification, focusing only on benefits associated with novel products. People who favour the use of the genetic modification are more likely to trust a source promoting its benefits. People who do not support the use of genetic modification in food production are more likely to distrust this same source providing information which does not align with strongly held views.

This does not explain why the same effect was observed in the case of the "balanced" information strategy. The reason may be because of the way in which the information strategies were developed in the first place – from the opinions of "experts" in the area of biotechnology, who proposed a rationalistic approach to technology communication issues. Expert views regarding what is salient to risk communication may be very different from what is considered important by the public. In order to assess whether this was, in fact, the case, it is useful to reflect on how the different information strategies were initially developed through the process of "stakeholder analysis". This will now be described.

Scholderer et al. (1999) conducted "expert focus groups" in order to understand the opinions of stakeholder groups regarding information dissemination about genetically modified foods. Technology experts believed that negative public attitudes resulted from a lack of information about genetically modified foods – specifically the lack of "objective" information was thought to cause uncertainty about their associated risks and benefits and, subsequently, negative evaluation of the entire technology. Thus the information strategies adopted reflected the views of the majority of experts, who appeared to be proponents of the so-called "deficit" model of risk communication, which assumes that public perceptions are inaccurate because they do not align with those of experts (Hilgartner 1990). This is discussed in greater depth in the next section. It is not surprising that the information produced was not perceived to be salient in discussing issues of direct concern to the public.

3.4 Institutional Denial of Uncertainty

Public negativity and resistance to food technology in Europe has been well documented, (Frewer 1999) and is paralleled by increased public concern associated with changes in production that are dependent on technological innovation or change. Genetic modification of foods is a case in point. Industrial and government concern about low levels of public acceptance of emerging technologies such as genetic modification of foods, the cloning of animals, or other advances in the biosciences has resulted in a communications industry growing in parallel with the industrial expansion linked to the growth of genetic technology in the agro-food sector. To some extent, the historical focus of this communication work has utilised the so-called "deficit model" (Hilgartner 1990). The deficit model promotes the idea that if only the public understood science, (or a simplified version of science) they would respond to technical risks in the same way as technical experts. Another version of the deficit model might assume that the public are unable to handle uncertainty information, and that, if the "deficit" is not amenable to public understanding through increased communication efforts, then information should be repressed or hidden to stop the public reacting in an "irrational" and "inappropriate" way.

There is indeed some evidence that expert groups do not believe that the public can handle information about uncertainty. Various food risk experts, drawn from scientific institutions, industry and government, were interviewed about how they thought the general public might handle information about uncertainty associated with food risk assessment (Frewer et al., in press). Many people within the scientific community expressed the view that the general public were unable to conceptualise uncertainties associated with risk management processes, and that providing the public with information about uncertainty would increase distrust in science and scientific institutions. These same experts also believed that uncertainty information would cause panic and confusion regarding the extent and potential impact of a particular hazard. This contrasted with the opinions expressed by the public (Kuznesof, submitted). A series of focus groups sampling members of the public drawn from different social milieus demonstrated that the general public were very familiar with the concept of uncertainty (perhaps through exposure to conflicting scientific opinion in the media, or through decision making experienced as part of everyday life).

The observation that scientists have a tendency to deny that the public can understand and handle scientific uncertainty has real world ramifications. This has been demonstrated by institutional responses to BSE (Bovine Spongiform Encephalopathy) in the UK. It is now known that the occurrence of BSE in cattle represents a serious health risk to humans, as well as having important consequences for the UK economy. At the time of writing, at least 80 human fatalities have arisen from new variant Creutzfeldt-Jakob Disease (vCJD). All these cases appear to be directly linked to the consumption of infected beef. In addition, the negative effects on the UK economy have been documented, although these were most acute during the high level of media reporting regarding the risk issues (Burton and Young 1997).

Prior to the announcement of the link between BSE and vCJD, government officials dealing with risk assessment information appeared to take the view that the public were unable to conceptualise scientific uncertainty, due to lack of insight and understanding regarding scientific processes, risk assessment and risk management. Scientific uncertainty was associated with the lack of knowledge regarding the minimum infective dose in the form of ingested infected material by both cattle and humans, and ignorance regarding the path of infection. It was thought that disclosure of the uncertainties linking the occurrence of BSE in cattle and vCJD in humans would produce public panic and alarm, accompanied by a public boycott of beef and beef products. This would have a detrimental effect on the UK economy. The public were, therefore, believed by scientific experts associated with the case to be "deficient" in their understanding of scientific process, and could not be trusted to respond in a "rational" way to the risks, which were deemed to be negligible and scientifically unsubstantiated. An alternative to the "rationalist" argument might be that, under conditions where there is uncertainty about the extent and nature of a risk, and where alternative behaviours to the potentially hazardous activity can be easily taken by those exposed to these uncertain risks, then it is actually quite rational to change behaviour to avoid the uncertainty. From this, one might argue that officials from the Ministry of Agriculture, Fisheries and Food were prepared to compromise public welfare in order to protect UK producers. This observation probably chimed with beliefs already held of many members of the public when the truth emerged (Frewer and Salter, submitted). From this, it is justifiable to postulate that it was not public "irrationality" that was the concern to the regulators, but rather the impact that a "rational" response by consumers would have on the UK economy. Whatever the underlying reasoning behind events, the BSE crisis not only had serious consequences for human health and the UK economy, but also for how the frameworks used for providing scientific advice to government are structured and operated.

The report resulting from the BSE inquiry (HM government 2001), identifies that events after March 1987 "demonstrated a policy of restricting dissemination of information about BSE", (p. 35) primarily because of the possible effects on exports and the political implications of discovery of the disease. The report also argues that concerns that a new Transmissible Spongiform Encephalopathy in Cattle would have a negative impact on the beef industry and export market did not justify suppression of information which would be needed if disease surveillance was to be effective, and there was to be early implementation of remedial measures. On page 233 of the report, it is stated that,

those concerned with handling BSE believed it posed no risk to humans (but) did not trust the public to adopt as sanguine an attitude. Ministers, officials and scientific advisory committees … were all apprehensive that the public would react irrationally to BSE. … the fear was that it would cause disproportionate alarm, would be seized on by the media and by some scientists as demonstrating that BSE was a danger to humans, and scientific investigation of risk should be open and transparent. In addition, both the advice and reasoning of advisory committees should be made public.

One result of increasing transparency in regulatory decision-making is opening up the uncertainties inherent in risk management to public scrutiny, which has direct implications for the communication of this uncertainty. Failure to discuss uncertainty with the public is likely to increase public distrust in risk management practices.

3.5
Public Characterisation of Food Risk Uncertainty

The communication of risk uncertainty with the public has been the focus of empirical investigation. Research has indicated that the public are able to articulate their views on uncertainty, are comfortable with the notion that uncertainty exists in food risk information and that uncertainty might be attributable to a variety of causes. The public also utilise their experience in dealing with uncertainty in food safety in the decision-making processes associated with new or emerging hazards (Frewer et al. 2001).

The majority view held by the public concerning uncertainty was that it was due to deficiencies in the present state of knowledge, for example through conflicting evidence or incomplete information. The minority view was that uncertainty arose due to the suppression of risk information. A further view was that the public viewed uncertainty as a transitory concept – the source of uncertainty was expected to be resolved over time through research or related activities. Admission of uncertainty also had a truthful, credible resonance or "high face validity" for the public, which appears to increase their trust in regulatory institutions. The degree to which uncertainty was regarded as acceptable varied substantially in relation to the cause of the uncertainty. The presence of uncertainty was found to be most acceptable when it originated through incomplete or conflicting evidence (i.e. limitations in knowledge) rather than the suppression of knowledge.

In general, the public expressed a strong preference for the provision of full information in situations when uncertainty arose with regard to food safety. Indeed, some people believed that it is their "right" to be informed about risk uncertainty associated with food. Respondents wanted information on how risk assessments were made as well as about the wider processes of risk analysis. Above all, people expressed a preference to be provided with the informational tools in order to make their own informed choice about food selection decisions, and to have the personal freedom to act upon that information. Food safety information was criticised for lack of openness, transparency and source credibility. It was concluded that best practice in risk communication should also consider the broad objective of providing meaningful messages to individuals to empower them to make informed choices about their personal food selection decisions. Information about food risk uncertainty needed to be accompanied by contextual information or a description of why the uncertainty exists, and address what information is needed to remedy the uncertainty. The contents of food risk messages should include the source of uncertainty, methods for remedying the situation, the foods affected, the most vulnerable groups at risk and the potential hazard posed.

In the UK, people exhibited similar preferences about the method of information delivery (explicitly preferring the television news, supermarket leaflets, and Government publications). Multiple delivery systems were the preferred way to deliver uncertainty information. In addition, people preferred the information to originate from Government and food industry sources, despite the fact that, in the UK, these sources are the most distrusted (Frewer et al. 1997). Taken together, the results mitigate against regulatory approaches which emphasise precaution over informed choice. All information about food risk uncertainties should be made available in the public domain, together with the means to enable consumers to make informed decisions (e.g. through an effective labelling policy).

3.6 What Drives Food Choices – Perceptions of Risk or Benefit?

Analytic approaches to decision-making may inappropriately treat perceptions of risk and of benefit as distinct concepts, although it is possible that the two concepts are not independent. Several studies have found an inverse relationship between perceived risk and benefit (Fischhoff et al. 1978; Alhakmi and Slovic 1994; Frewer et al. 1998). Indeed, Alkhami and Slovic (1994) have observed that, if this inverse relationship holds, it may be possible to change perceptions of risk by changing perceptions of benefit, and vice versa. Thus for a hazard which people perceive to be high in risk and low in benefit, reducing risk perceptions may be brought about by increasing perceptions of benefit rather than heightening perceptions of safety. Empirical support for this premise is weak (Frewer et al. 1998), although variation may be associated with the extent to which attitudes have crystallized and are amenable to influence by new information (Frewer et al. 1999). The relationship may also depend on the extent to which benefits specific to a potential hazard are perceived to be desirable, or the associated risks intolerable.

There is substantial evidence that, just as psychological constructs associated with risk may be very specific to the type of hazard under consideration, so may perceptions of benefit (Miles and Frewer 2001). It is important to understand what members of the public perceive to be benefits, as opposed to what is believed to be beneficial by technical risk experts. Misunderstanding public concerns can have very negative consequences for effective risk communication. Information must be relevant and important to consumers if they are to read it and think about the contents in an in-depth way. This is the basis on which consumers are able to make informed choices, and without which arguments that the public implicitly approve the technology cannot be made. If, for example, consumers are very concerned about ethical issues, and the information does not address these, relevance is reduced and the information is not read or thought about (Frewer et al. 1999). In particular, there has been considerable political pressure from the scientific community to focus communication efforts on the issue of substantial equivalence, which is assumed to imply that genetically modified foods are unlikely to effect human health, thus facilitating public acceptance (FAO/WHO report 2000). However, there is evidence that the public are concerned about environmental impact or other process-related risks. The communication is, unsurprisingly, at best likely to appear irrelevant, or at worst be perceived by the public as an attempt to hide the "real" risks of the technology from them in order to promote internal and opaque scientific or industrial agendas.

Foreman (1990) has noted that emerging technologies may result in public resistance if the resulting risks and benefits do not accrue equally between different groups within the population. For example, if the public believe that the benefits of the technology apply only to industry, or other stakeholders, but the risks will impact on the environment and affect the whole population, then one might predict a negative public response. This type of effect may extend to other hazard domains as well as those linked to emerging technologies. Perhaps of even greater concern is the perceived differential accrual of risk to specific demographic or geographic groups within the population, particularly if these groups perceive themselves to gain no benefit from hazard exposure and to be excluded from risk management decision processes (Frewer 1999). One solution may be to increase public consultation and public participation in risk management, so that the decision-making process is believed to be equitable and fair.

Many different types of public participation methodology have been identified in the literature (e.g. Fiorino 1990; Renn 1995). These range from those which elicit input in the form of opinions (e.g. public opinion surveys and focus groups) to those that elicit judgments and decisions from which actual policy might be derived, and which are essentially deliberative in nature (e.g. consensus conferences and citizens' juries). Space does not permit a substantive review of the different methodologies, and the interested reader is referred to Rowe and Frewer (2000) for a more detailed review of methodological approaches in this area. However, it is interesting to note that the practice of public participation has increased across all areas of policy development in recent years, although issues of "best practice" are disputed. Such procedures, which aim to consult and involve the public in decision making, include diverse methodologies, ranging from traditional opinion polls (low in deliberative input from participants) to public hearings, referenda, focus groups, consensus conferences and citizens juries (high in deliberative input). The success of these different methods has been measured in an ad hoc way, if measured at all. Rowe and Frewer (1999) have specified some theoretical criteria for benchmarking the effectiveness of public participation exercises, which are currently being tested in real world contexts. Broadly speaking, evaluative criteria fall into one of two categories: those related to public acceptance of a procedure (that is, "Acceptance Criteria"), and those related to the effective construction and implementation of a procedure, which refer to the procedural issues associated with the participation exercise itself ("Process Criteria"). These criteria, and the process of validation of these criteria, are described in greater detail elsewhere (Rowe and Frewer 2000; Frewer, in press). However, the potential effectiveness of public consultation may be compromised by failure to evaluate not only the process but also the substantive impact of the process on policy. Frewer and Salter (submitted) have argued that recommendations for best practice regarding public consultation and public involvement must include the explicit assessment of both scientific advice and public consultation on policy development if public confidence in science and risk management is not to be further eroded (Frewer 1999).

3.7 Conclusions

Initial research into risk perception was conducted in order to understand why the public did not react to potential hazards in a "rational" way that reflected risk mitigation priorities as defined by probabilistic analysis. The "psychometric paradigm" was developed as a basis for understanding people's risk perceptions, and identified that risk perceptions were driven by factors other than technical risk estimates. In contrast to earlier assumptions, the psychometric paradigm has demonstrated that people judge "risk" in terms of dimensions other than probability and harm: control, the extent to which exposure to a hazard is voluntary, dread, and catastrophic potential are all important determinants of people's anxieties concerning food choices and, indeed, hazards in general. The initial understanding of what was driving public risk perceptions developed theories focused on effective risk communication. Although some authors have argued that theoretical developments in risk communication were geared towards technology acceptance, this is not the case in the food area, where communication issues range from microbiological risk, (where communication efforts tend to focus on establishing good domestic hygiene practice) or crisis management (for example in the case of BSE or contamination of food by dioxins). Whilst 10 years ago the emphasis of communication about the use of technology in the food chain was geared towards technology acceptance, the emphasis is now on increased public involvement in deciding how to manage and regulate technology innovation. In particular, there has been increased stress in recent times on communicating information relevant to people's concerns, as well as conveying information about probabilistic risk assessment processes. The failure of the Richter scale approach to risk communication (where relative risk probabilities are explained in a simple, often diagrammatic, approach) is thus explained. Comparative risk communication is driven by comparison of technical risk estimates that do not take account of psychological factors that contribute to risk characterisation, although people need information beyond technical risk estimates (Adams 1997).

More recent research has implied that trust in science and risk regulators, and public confidence in scientific advice, has powerful explanatory power in the context of how people respond to and interpret information. Recent theoretical stances have developed the idea that distrust of institutions (partly through perceived exclusion from the decision making machinery linked to government and science) represents a key driver in creating and fuelling public negativity to scientific innovation and risk management practices.

Efforts to understand the psychological determinants of trust (in information sources and regulatory institutions) laid the groundwork for subsequent analysis of how complex risk information is processed and transmitted by individuals. However, the need for explicit public involvement in risk management policy has emerged as a key driver in initiatives to increase public confidence in risk management. It is arguable that a weak aspect of increased public consultation lies in the way in which outputs of consultations are explicitly used in policy development in the long term and evaluative procedures to assess the impact of these outputs on policy must be developed. It is increasingly clear that simply telling people about different risks will neither reduce their cynicism regarding risk management practices nor the quality of scientific advice, unless the information is relevant, perceived to be truthful, and is honest about the uncertainties inherent in risk analysis.

References

Adams J (1997) A Richter Scale for risk? The scientific management of uncertainty versus the management of scientific uncertainty. In: Vitale M (ed) Science and technology awareness in Europe: new insights. The European Commission, Brussels, pp 93–111

Alhakmi AS, Slovic P (1994) A psychological study of the inverse relationships between perceived risk and perceived benefit. Risk Analysis 14: 1085–1096

Barke RP, Jenkins-Smith HC (1993) Politics and scientific expertise: Scientists, risk perception, and nuclear waste policy. Risk Analysis 13: 425–439

Burton M, Young T (1997) Measuring meat consumers' response to the perceived risks of BSE in Great Britain. Risk Decision and Policy 2: 19–28

Carmody P, Woolridge M, Warburton D, Frewer LJ (submitted 2001) Exploring peoples' beliefs and about organophosphate sheep dip using a qualitative approach: Impact of group membership

Cvetkovich G, Löfstedt RE (1999). Social trust and the management of risk. Earthscan Publications Ltd, London

Da Costa MC, Deliza R, Rosenthal A, Hedderley D, Frewer LJ (2001) Non-conventional technologies and impact on consumer behaviour. Trends in Food Science and Technology 11: 188–193

Deliza R, Rosenthal A Hedderley D, MacFie HJH, Frewer LJ (1999) The importance of brand, product information and manufacturing process in the development of novel environmentally friendly vegetable oils. Journal of International Food and Agribusiness Marketing 10: 67–79

FAO/WHO Joint Expert Consultation on Foods Derived from Biotechnology, Geneva, 29 May to 2 June 2000. http://www.who.int/fsf/GMfood/Consultation_May2000/Documents_list.htm

Fischhoff B, Slovic P, Lichtenstein S, Read S, Combs B (1978) How safe is safe enough? A psychometric study of attitudes towards technological risks and benefits. Policy Sciences 9: 127–152

Flynn J, Slovic P, Mertz CK (1993) Decidedly different: Expert and public views of risks from a radioactive waste repository. Risk Analysis 13: 643–648

Frewer LJ (1999) Risk perception, social trust, and public participation into strategic decision-making – implications for emerging technologies. Ambio 28: 569–574

Frewer LJ, Salter B (submitted 2001) Public attitudes, scientific advice and the politics of regulatory policy: the case of BSE

Frewer LJ, Howard C, Hedderley D, Shepherd R (1996) What determines trust in information about food-related risks? Underlying psychological constructs. Risk Analysis 16: 473–486

Frewer LJ, Howard C, Shepherd R (1997) Public concerns about general and specific applications of genetic engineering: Risk, benefit and ethics. Science, Technology and Human Values 22: 98–124

Frewer LJ, Howard C, Shepherd R (1998) Understanding public attitudes to technology. Journal of Risk Research 1: 221–237

Frewer LJ, Hunt S, Kuznesof S, Brennon M, Ness M, Ritson R (in press 2002) The views of scientific experts on how the public conceptualise uncertainty. Journal of Risk Research

Frewer LJ, Hunt S, Miles S, Brennan M, Kuznesof S, Ness M, Ritson C (2001) Communicating risk uncertainty with the public. Final project report, February 2001. University of Newcastle, Newcastle

Frewer LJ, Scholderer J, Bredahl L (submitted 2001) Communicating about the risks and benefits of genetically modified foods: Effects of different information strategies

HM Government (2001) The interim response to the report of the BSE Inquiry by HM Government in Consultation with the devolved Administrations. The Stationary Office, London

Hilgartner S (1990) The dominant view of popularisation: conceptual problems, political uses. Social Studies of Science 20: 519–539

Hovland CI, Janis IL, Kelley HH (1953) Communication and persuasion: Psychological models of opinion change. Yale University Press, New Haven, CT

Hunt S, Frewer LJ, Shepherd R (1999) Public Trust in Sources of Information about Radiation Risks in the UK. Journal of Risk Research 2: 167–181

Johnson B (1999) Exploring dimensionality in the origins of hazard-related trust. Journal of Risk Research 2: 325–354

Kuznesof S (2001) Understanding lay conceptualisations of scientific uncertainty. Report to the UK Food Standards Agency. University of Newcastle, Newcastle

Kraus N, Malmfors T, Slovic P (1992) Intuitive toxicology: Expert and lay judgments of chemical risks, Risk Analysis 12: 215–232

Larkin JH (1983) The Role of problem representation in physics. In: Gentner G, Stevens AL (eds) Mental Models. Lawrence Erlbaum Associates, London

Lazo JK, Kinnell JC, Fisher A (2000) Expert and layperson perceptions of ecosystem risk. Risk Analysis 20: 179–193

Mertz CK, Slovic P, Purchase IFH (1998) Judgements of chemical risks: Comparisons among senior managers, toxicologists, and the public. Risk Analysis 18: 391–404

Miles S, Frewer LJ (2001) Investigating specific concerns about different food hazards – higher and lower order attributes. Food Quality and Preference 12: 47–61

Rowe G, Wright G (2001) Differences in expert and lay judgments of risk: Myth or reality? Risk Analysis 21: 341–356

Scholderer J, Balderjahn I, Bredahl L, Grunert KG (1999) The perceived risks and benefits of genetically modified food products: Experts versus consumers. European Advances in Consumer Research 4: 123–129

Slovic P (1993) Perceived risk, trust and democracy. Risk Analysis 13: 675–682

Siegrist M (1999) A causal model explaining the perception and acceptance of gene technology. Journal of Applied Social Psychology 29: 2093–2106

Sjöberg et al (2000) Through a glass darkly: Experts' and the public's mutual risk perception. Foresight and precaution. Volume 1. Cottam MP, Harvey DW, Pape RP, Tait J. Rotterdam, AA Balkema: 1157–1162

Sparks P, Shepherd R (1994) Public perceptions of the potential hazards associated with food production and food consumption: An empirical study. Risk Analysis 14: 799–806

Woolgar S (1996) Psychology, qualitative methods and the ideas of science. In: Richardson STE (ed) Handbook of Qualitative Research Methods. British Psychological Society, Leicester, pp 11–24

Food Risks, Public Policy and the Mass Media

4

JACQUIE REILLY

4.1 Introduction

The communciation of risk information has become an important concept in contemporary public and political debate and the mass media are seen to play a key role in this social transformation. Pressure groups seek to attract media attention in their campaigns for safety measures, experts complain of media 'scare-mongering', industries and government bodies employ special 'risk communicators' in an attempt to gain or maintain public confidence, and journalists themselves describe the attractions of scientific controversy and risk disputes (Adams 1992; Friedman et al. 1986; Hansen 1994, Sandman 1988).

Several studies suggest that the media are paying increasing attention to scientific uncertainty and have been instrumental in generating public concern about particular threats (Cole cited in Goodell 1987; Goodell 1987; Peters 1995). However, media reporting has also been shown to emphasise risk in favour of offering reassurance (Schanne and Meier 1992). Furthermore, some research argues that the media have failed to address fundamental dangers integral to corporate practice, uncritically presented technological developments as 'progress', overlooked severe threats to public health and only belatedly attended to crucial social problems brought into the public domain via other avenues (Banks and Tankel 1990; Medsger cited in Goodell 1987; Schoenfeld et al. 1990).

At the very least it is obvious that media coverage of risk is selective. Even if risk is inherently newsworthy, not all risks can be in the news all of the time. Research also consistently demonstrates that the media do not simply reflect experts' assessments of the most 'serious' risks and reporting does not parallel incidence figures. For example, the distribution of media reports about incidents of Salmonella poisoning diverge completely from the amount of media attention this form of food poisoning attracts (Miller and Reilly 1995).

In order to understand media reporting of food risks we need to examine the process by which a risk emerges in the mass media. Far from being eager reporters of risk, the press and TV news are ill-adapted for sustaining high-level coverage of long-term threats. Media interest is rarely maintained in the face of on-going scientific uncertainty and official silence or inaction. In spite of this, pressure groups can use the media to force an issue onto the public agenda in the face of official denials; and the media can serve as one avenue for public information and political/policy leverage for those who believe that risk assessment is 'too important to leave to the experts'. However, the media cannot be assumed to be automatic allies in the 'democratisation of risk'.

4.2
The Media as a 'Black Box'

Although the mass media are perceived as being important in the area of risk communication research fails to explore how the battle for the definition and validity of different risks is fought out. The mass media are not simply reflecting a 'new epoch', nor are they indiscriminately attracted to risk. In fact, the mainstream news media are ill-designed to maintain attention to any particular future threat. Individual stories will attract attention when there are decisive scientific statements, major disasters, fresh human interest stories, official reactions and/or when major organisations or governments come into conflict over the extent of the danger. However, many risk debates, much of the time, do not fall into these categories and, regardless of the concern of certain journalists, news structures do not encourage sustained risk coverage. There are a number of reasons for this.

Firstly, risk is often characterised by uncertainty and the absence of conclusive scientific evidence ('virtual risk', Adams 1995). This leads to 'going round in circles' and the 'we need more research' approach which is so frustrating for journalists. Scientific uncertainty per se, is *not* attractive to journalists. New and apparently definitive findings and controversy are more likely to draw media attention.

Secondly, unless governments, or other official bodies, adopt the precautionary principle, risks may be ignored or appear to be resolved at the official level which, in turn, dampens the story's news value. The irony here is that it is precisely this failure to adopt precautions which may increase risk.

Thirdly, and perhaps most important of all, risk, by definition, concerns projected assessments. Risk is a concept based on predicting the future. This is in conflict with the basic news principle that emphasises the events of the day. Press and television news, far from focusing on risk, may actually tend to ignore very distant and hypothetical threats. Many potential risks will not be reported as risk stories unless or until the dangers are manifest in some way. The news media are better at retrospective than prospective reporting of risk, and retrospective risk reporting is inherently limited.

Official accounts of expert assessments and their transformation into policy tend to present these as based on science, logic and systematic rational or bureaucratic procedures. If the system runs smoothly, policy making, it is argued, is based on the best available expert evidence. The mass media may be seen either to reflect the 'truth' or (more often) to distort it, and public opinion may be seen as an important site of consultation or as a problem to be handled.

A sociological analysis of these processes highlights a different perspective. It would suggest that the media are involved throughout the risk management process and are involved in a battle for attention at every level: from establishing the availability and definition of the best expert advice through to policy decisions, and their representation to the public.

This chapter will attempt to account for the way a particular risk is represented in the media. The chapter is based on research which looked at how the mass media may help to construct risks within the public arena.[1] One of the substantive areas studied was the issue of BSE. (Kitzinger and Reilly, 1997). The BSE story has perhaps been the most dramatic 'risk crisis' in Britain so far this decade. It has been an extremely long running and complex saga with a definite risk narrative. The existence of this new disease, Bovine Spongiform Encephalopathy (BSE) was first announced by Ministry of Agriculture, Fisheries and Food (MAFF) in 1987. In April 1988 the British government set up a committee (Southwood) to assess its significance. Southwood reported back in February 1989. On the potential danger of BSE to humans, the report concluded that it was: "most unlikely that BSE will have any implications for human health. Nevertheless, if our assessments of these likelihood's are incorrect, the implications would be extremely serious". It also stated that: "With the long incubation period of Spongiform Encephalopathies in humans, it may be a decade or more before complete reassurances can be given" (Southwood 1989). So, while scientific knowledge had not ruled out the possibility that BSE could be transmitted to humans, the Ministry of Agriculture, Fisheries and Food stated that the report's findings meant that there was no risk to humans and beef was safe. This position did not change until March 1996.

The handling of the BSE issue and its implications for human health has been subject to controversy throughout its history. This issue involves 'risks' associated with modern agricultural policy, the effects of intensive farming, the threat of zoonoses, and the role of government in ensuring the safety of animal and human health. BSE raises questions about how scientific uncertainty was translated into risk judgements, and policy decisions, who influences those decisions and how. What we want to look at here, from a sociological

[1] The project was entitled 'Media and Expert Constructions of Risk' and was funded under the ESRC's Risk and Human Behaviour Programme.

point of view, are the processes by which risk assessments were made, in relation to BSE, and the role of the mass media in the communication process. The following sections will look at a number of important issues which were developed in the original research project. The use of some data such as graphs and quotations from those interviewed in the project will be used to highlight points.

4.3 How Risk Assessments are Made

Risk assessment often involves exceptional levels of uncertainty. Scientific experts who were interviewed made comments such as, "*No-one knew anything about it [BSE].*" But, assessments are likely to change as new evidence emerges. However, until some resolution is achieved, risk disputes often involve a breakdown of professional consensus, the overt questioning of assumptions, and a need to step beyond existing areas of expertise. These disputes often raise questions about who are the 'experts', what counts as evidence, and how it should be interpreted. Indeed, debate about how truth is assessed lies at the heart of the controversy surrounding BSE which has been characterised by fierce arguments about the interpretation of findings. Risk crises thus expose divergent risk assessment or presentation processes. A number of factors can influence processes and opportunities for risk assessment.

4.4 Motivation to Assess Risk

The allocation of resources such as funding and time to risk assessment depends on political/professional/commercial motivation. The type of research conducted is influenced by the remit of funders/employers and by access to data. In some cases access may be limited because of 'commercial confidentiality' and/or the actions of government.

4.5 Frameworks

Conceptual framing influences which questions/evidence are considered relevant. For example, using sheep scrapie as the disease model for BSE and viewing it as an animal health problem created a different framework for risk assessment than seeing it as a potential public health issue. In addition, different types of data are generated by different techniques (such as injecting different species with BSE infected material versus, for example, mathematical modelling). Each set of findings may then be subject to diverse interpretation.

4.6 Sources

Expert assessments often rely on published research or other secondary sources. Source credibility is therefore crucial. This can operate both on a general personal/political, as well as scientific, level. One challenger to official government pronouncements about the risks from BSE given a lot of media coverage initially was labelled on various occasions as 'a charlatan' and a 'sensationalist' by official sources. But, such dissenting voices, such as Richard Lacey's, gradually lost the power to attract media attention and became isolated. As one journalist observed: *It was largely seen to be one man crying wolf. ...you think...'Could this man be the Copernicus of his field or is he a nut case?' ...(Copernicus) might well be the Richard Lacey of his field. I can imagine* ***The Medieval Times*** *of the day saying: 'This man, Copernicus, should be burnt at the stake. He's obviously a nut case'...No matter how provocative, however interesting the theory, unless the rest of the scientific community catch up, this person is out on a limb" (Broadsheet specialist). Without any new data to support him, Lacey's views soon became 'old news': "One of the most devastating things a news desk can ever say to a journalist is: 'I think we know that, don't we?' in a very sarcastic voice. So it wouldn't have done me any good to wander up and say 'Oh, Professor Lacey is still very worried that we might all get some weird brain disease from moo cows'. The news desk would say: 'Sod off, we know that, he's said it before'* (Broadsheet specialist).

4.7 Judgements about Risks

Experts' judgements about risk do not simply depend on assessing the possibility of a danger, but also on its probability and thus on ideas about how society operates. For example, MAFF's caution over what could be said in public minimised the chance that official sources would make controversial statements. For example, high-profile official experts who said in interview that they personally were taking precautions, were not prepared to say that in public. As one official expert commented:

...as far as we know at the moment there doesn't seem to be a risk to humans. While I personally don't believe that there ever will be, we just don't know at the minute, but that's science for you.' [JR: 'So, do you eat bovine offals at all?] 'Well, no I don't actually. But I could never say that in public because I shudder to think what the media would make of it. And I don't think that it would be very sensible professionally given the highly emotive nature of the subject. (Official expert interviewed in 1995)

4.8
Professional Assessment Versus Personal Reactions

Professional risk assessments can differ from personal assessments and behaviour. Most notably 15 of the 'BSE experts' who, at the time of interview, believed (and often stated in public) that beef was, as far as they knew, completely safe, had stopped eating some beef products themselves.
The transformation of expert assessments into policy decisions involves the motivation to formulate policy, selection of which experts to involve, and policy makers' presentation of, and response to, expert recommendations.

4.9
The Motivation to Make Policy

Decisions to formulate policy depend on perceiving that a problem exists or that there is media/public/international concern. The form and level of policy reaction is then related to where responsibility is seen to lie, a process which is a subject of struggle, and influenced by political philosophy or policy objectives.

4.10
Channels of Communication

Once policy is seen to be necessary, then which expert advice is fed into policy making processes depends partly on the definition of 'relevant' experts and established channels of communication. One key line of communication is between high-status professional organisations and government. A second involves the expert committee or professional working party, for which the selection procedure is itself usually opaque and sometimes controversial. Formal processes, however, are not the only way in which experts influence policy. Experts in the areas we studied were also invited to contribute to policy discussions as a result of their pressure group activity or fed into the process via the media and direct action. For example, an Environmental

Health Office document was sent to MAFF in February 1990 pointing out poor practices in abattoirs. No reply was received, so as a representative of The Institute of Environmental Health Officers explained: "We approached certain journalists and said: 'look we've found out that there are some disgracefully risky things going on in abattoirs, and something has to be done about it. Will you print it?'...With government it is necessary to get the ball rolling, everything takes such a long time. But if there is public concern that can move things along...It's not the ideal way of doing things, but when needs must."

4.11 The Presentation of 'Expert' Assessments

Civil servants draft and revise a report from an expert committee with a view to what ministers will accept. Subsequently, its translation into public statements may involve modification whereby caveats included by experts are excluded (see MAFF/DoH press conference on Southwood committee report into BSE). Similarly, papers published in scientific journals are used in the lobbying process and may be employed to vindicate policy in ways which the editors of those journals consider inappropriate.

4.12 The Policy Response

Risk disputes usually involve scientific uncertainty and the same expert assessment can lead to either a precautionary or wait-and-see response. Policy decisions therefore necessarily go beyond the purely scientific and involve other factors. For example, the formation of BSE policies may have been influenced by:

a) History: Previous crises such as Salmonella in eggs (1988/89) informed response. As one DoH civil servant commented: *"We learnt that we had to be ultra careful about what people said. Word came from the top that care had to be taken in all aspects of the job. There was no way another fiasco was going to be allowed to happen."*

b) Allocation of 'the burden of proof': The 'burden of proof' was placed on those who argued that there might be implications for human health. As a DoH official commented: *"Everyone at that political level was concerned not to damage beef sales, exports etc. ...the paramount concern was...not to say anything without 100% proof which could be seen as endangering that. This of course was the problem, how can you get that proof? Even if the agent could be*

passed into mice, that was never going to be enough to say that it would go into humans."
c) Balancing risks, the allocation of responsibility: Policy makers are involved in 'balancing risk' (e.g. economic risks versus health risk). The definition/parameters of the problem and departmental allocation of responsibility are crucial. Policy responses to BSE were, at some level, determined by MAFF and influenced by its agricultural remit.

The processes outlined above are usually invisible in journalists' reports, but may become the subject of media enquiry once an issue is defined as a 'risk controversy'. However, this definition is not an automatic reflection of 'fact'; and some risk controversies are entirely ignored by the mass media.

But BSE involved a dramatic risk crisis with coverage across every media outlet and format. Its profile is distinguished by a peak of mass media concern in 1990, followed by a relative lull until 1996 when it became a major risk story again. As with other issues it is clear that the level of media coverage given to any given issue does not necessarily mirror the incidence of the disease/problem in reality (see figure 1 below).

A number of factors affected media attention and influenced the profile of BSE.

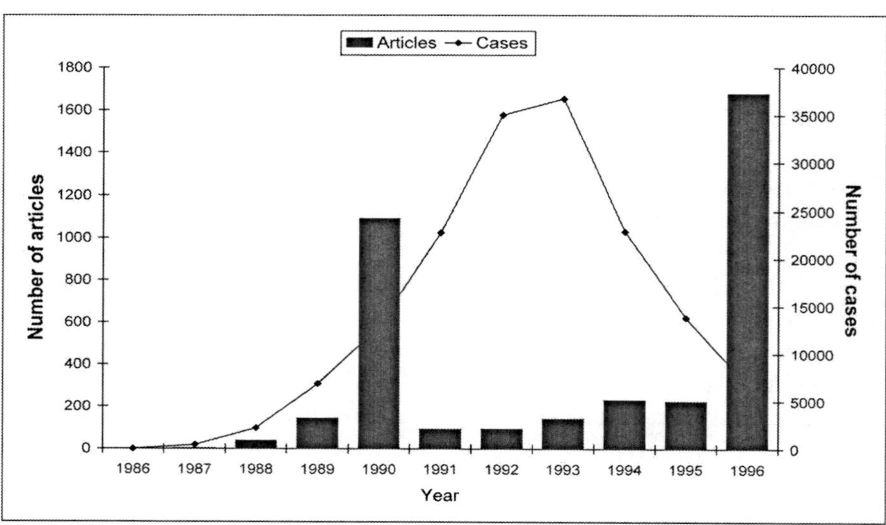

Figure 4.1 BSE: number of cases of infection in the UK against the number of articles in UK national newspapers (1986–1996)

4.13 Journalists' Training

Journalists' willingness to explore a specific risk is influenced by their confidence in that area. This is partly a question of training, but also of their perceptions of their knowledge. Some clearly took on a campaigning role. It was David Brown of the *Daily Telegraph* who is credited with coining the phrase 'Mad Cow disease'. While the name, unsurprisingly, '*wasn't the favourite of the agricultural industry*' he claims that giving it such a catchy title was, in his words, a '*public service*': He said that at the time:

The NFU did't think that it was a particularly good thing to do but at the same time they couldn't get anyone interested in how the farming industry was going to fight the disease. They knew it had to be stopped, they knew it was serious and where was the money? The treasury is only going to produce 50 % of the money needed, how do you persuade the treasury that you get the other 50 %? There is only one way and that's through public pressure, through MP's... there is no public perception of urgency... the title of the disease summed it up. It actually did a service. I have no conscience about calling it mad cow disease....

At another level some journalists felt that once the political aspects of the issue were perceived to have been cleared up (lifting of European bans in 1990) the uncertainty and complexity of the science did not make 'good copy'.

4.14 News Values, 'Fresh News' and Policy Announcements

The lack of coverage of BSE following 1990 was partly explained by the fact that certain risks were seen as very distant, they were, one journalist commented: '*speculation not news*'. The seeming lack of policy activity also undermined media interest. The media profile of BSE was also directly influenced by political events. BSE had a great deal of news momentum during 1990 because of the level of policy/official activity, but this attention declined when there appeared to be a political resolution to the problem. Thus there were 1092 newspaper items about BSE in 1990, but only 93 in the following year. Journalists who maintained a personal interest during this time were thwarted by editorial decision-making:

They [the editors] lost interest in the subject because nothing was happening. Of course that was the whole point, nothing was happening to destroy this thing, but in newspaper terms I wouldn't be given the space to say that every day or every week..., what had to happen for the full scale go ahead of a major story was dead people.... (Broadsheet journalist)

4.15
'Human Interest'

BSE was identified by journalists as directly relevant both to themselves and to their readers, the view was that any of us could be infected by eating beef. At times, the topic provided human interest stories in the form of individuals who appeared as 'living proof' of the dangers (e.g. film footage of people apparently dying of CJD (Creutzfeldt-Jakob disease) helped to maintain the momentum of media coverage in 1996).

4.16
Media Templates

Coverage is influenced by 'media templates'. The media attention given to BSE, in part, built on a history of concern about food management. For example, there was already a well developed interest in food safety because of salmonella and listeria which were high profile public issues throughout 1988/9. Secondly, by 1989, other countries began to be interested in the disease (Australia had already banned British beef cattle exports in July 1988). Germany, Italy and France banned British beef imports. The issue, in British terms, became political and economic. European countries claimed they were protecting the public health. John Gummer treated this as powerful vested interests playing at protectionism. Thirdly, in Britain, local councils began banning British beef from the menus of 2000 schools. The death of a domestic cat from a Spongiform Encephalopathy caused alarm, opening the debate on transmission and bringing the potential threat to humans a little closer to home. As one food Industry representative put it: "*Everything would have died down had it not been for that bloody siamese cat*".

The media's initial attraction to BSE is easy to explain in the light of the factors already identified. The potential implications for human health were obviously relevant to the general public. Secondly, media interest was already primed to food crises stories, so it could be reported as further evidence of a crisis in the management of food risks and another reason to distrust government policy. Thirdly, 1990 saw a rapid build up of both self-referential and interactive news momentum; and finally, although BSE could not, at first, offer stories around human transmission, it could introduce individual case studies such as 'the ruined farmer'. It also had the added appeal of 'good pictures': Daisy the staggering mad cow, the Minister for Agriculture feeding his daughter a beefburger and the dramatic image of burning cattle carcasses. (TV news reporters, in particular, stress the importance of good film footage to ensure coverage, see also Jacobs 1996).

Crucially, the spread of a new, mysterious and rapidly increasing cattle disease had huge political/economic as well as potential health implications. It also involved clear government responsibilities and responses. Journalistic interest in the story was increased by the behaviour of the main government department involved in the crisis: the Ministry of Agriculture, Fisheries and Food.

At the centre of this whole issue is the role of science. Had more been known about the BSE agent, clearer statements about diagnosis and treatment could have been made. But, what became clear quickly was that until **scientific** uncertainties about mad cow disease were cleared up, reassurances about the safety of British beef were not entirely convincing, and no firm resolution to the problem could be reached.

While no-one believed that the government wanted to infect the population with a potentially fatal disease, decisions on BSE could not have been made based on pure science alone. Because of the lack of science what should have been a scientific debate opened up into an economic and political one within MAFF. BSE remained in the public sphere because controversy surrounded the subject.

Ultimately, although all the above factors contributed to media attention it was socio-political events which led to the original rise of the BSE story. These are traditionally events with high news values, especially the European dimension. In addition, this range of circumstances mobilised science, education, farming, food, political and economic correspondents, rapidly increasing the space into which BSE news could be inserted.

The importance of such events becomes even clearer when tracing the **decline** in media interest in BSE. While the first peak of media interest appeared in 1990, interest rapidly declined toward the end of that year and the next major peak did not occur until 1996. Given recent events, one of the most fascinating questions is: why did media attention lapse between 1990 and 1996 (when very clearly the disease itself was not in decline).

The fading interest had nothing to do with a change in MAFF's activity, nor any kind of scientific resolution, nor a decline in the spread of BSE. Rather, coverage reduced because there was a resolution of sorts on a political level. A compromise solution was instituted that reinstated beef imports to Europe so long as they were certified to come from BSE free herds. However, as we argued in 1995: "The central issue of human transmission was not resolved by the European community decision. Indeed, it can be argued that the certification of British beef exports to Europe made the official position less credible" (Miller and Reilly 1995, p. 322).

In spite, or perhaps because, of the outstanding scientific uncertainty, the decline in coverage toward the end of 1990 was sustained for the next 5 years, with only the occasional minor peak. BSE retained the potential to re-emerge but required further scientific evidence or renewed official action. At this stage, MAFF's efforts to control the story were actually quite effective.

The general low level of media interest in Britain during 1991–1995 was due, in part to the feeling that BSE had exhausted its news value. There were also very few new events to maintain a momentum of media concern. While a number of journalists remained intensely interested in BSE, they fell foul of editorial decision-making and the demands of hard news. One specialist correspondent commented: *"Scientists continually said 'we don't have the data, we need further research.'...so we tended not to write about it...It just doesn't make very good copy, to simply say 'we don't know, we need further research', 'we can't answer that'. However honest that is, it doesn't play very well in terms of headlines"* (Broadsheet specialist).

At the same time the lack of policy activity meant that editors lost interest in the subject because 'nothing was happening'. This reaction frustrated some journalists, because, as one broadsheet specialist pointed out: *"Of course that was the whole point,* ***nothing was happening*** *to destroy this thing; but in newspaper terms I wouldn't be given the space to say that every day or every week".*

The period 1991–1995 thus saw a relative lull in coverage. The 'information vacuum' created by MAFF still allowed for some dissenting voices to be heard occasionally, and much lobbying still focused on the media in an attempt to bring about changes in practice (e.g. in relation to abattoirs). However, blips which did occur, in the main, were linked to high news value events rather than the successful strategies of dissenting sources (for example, during 1992 further action from Europe again caused a brief flurry of reports and, in 1993 there were some suspected new cases of CJD). BSE did not capture the headlines again until March 1996 and then the change was very dramatic indeed.

By the end of 1995 there were 10 new cases of CJD which had appeared in younger people. These cases were similar to each other both in clinical symptoms and in the pathological damage that appeared in the brain (in similar areas to those found in cows with BSE). John Pattison, chair of the Spongiform Encephalopathy Advisory Committee (SEAC) suggested that projected cases of BSE in humans, calculated on current information, could represent a major public health problem. Under Pattison's head-ship SEAC decided that the news had to be made public and this led to an explosion of media coverage, even exceeding the previous peak of interest in 1990. Not only did BSE have all the high news values it had in 1990 but this time health interests were finally brought into play. In addition, the government cover-up/failure angle was all the stronger and European interventions were dramatically strengthened in 1996 with demands for a major culling policy. This time around too, the risk was less hypothetical and the case studies of individuals apparently dying of CJD were there for the cameras. As one journalist said, before his news desk would countenance giving a lot of attention to BSE: "*We needed dead people, well, we've got them now*" (Broadsheet specialist).

4.17 'Public Opinion'

The media respond to the assumed opinions of their target audience, depending on factors such as those outlined above and on specific page or newspaper identity. The desire to be seen to mobilise public opinion or prompt investigations and policy responses can also influence coverage, especially in some formats such as TV documentaries, encouraging strong single messages highlighting risk. Journalists also respond to 'public opinion' as represented by other media output or by colleagues. This creates a self-referential media momentum, whereby media coverage generates further attention in an upward spiral, eventually leading to 'news fatigue'.
Within the above context, a number of additional factors can be seen to influence who is heard, which views are reported, how and when:

4.18 The Nature of the Assessments

On the one hand journalists work with notions of balance along a continuum of 'reasonable' opinion. On the other hand uncertainty is less news-worthy than certainty and 'moderate' opinions less attractive than 'extreme' points of view. Experts also reported straightforward misrepresentation of their findings in order to support a particular story-line and some, very experienced, public commentators remarked that they had never had such problematic contact with the media before. It seems that when journalists cover risk stories they are more likely to challenge (or distort) experts' views, not least because the definition of 'expertise' itself comes into question.

4.19 Story Initiation, Timing and Source Resources

Official experts have the status to initiate news-worthy events, in particular, press conferences. This allows for a high degree of news control. Careful briefing, pressure of deadlines and official status can be used to maximise the chance that official assessments will be transmitted uncontested. However, in some cases, lay assessments of risk may enter the media far in advance of organised professional response. Although lay pressure groups may lack authority and material assets, they may be able to respond quickly and use other resources (including their human interest value) to mobilise publicity.

4.20 Pressures not to Talk to the Media

Some experts withheld their opinions from the media, because they themselves resisted exposure or, in some cases, they felt pressurised or were directly prohibited from expressing them. One BSE expert commented:

I've had criticism from my own health authority about getting involved in political aspects…it wasn't for me as what was described as a 'back-room researcher' to be commenting on these things of national importance.

4.21 Source Credibility

In addition to the power of official bodies to set up 'news events' and influence what was said in public, journalists are predisposed to accord official sources (e.g. those on government committees) greater credibility than unofficial or lay sources. Journalists are also more likely to rely on 'mainstream' professionals than 'lone voices' and are influenced by the generic status of particular professions (e.g. the relatively high status of doctors and scientists versus the dubious status of therapists and psychologists). However, this intersects with the perceived independence of the experts and their personal credibility. Crucially, once a story is defined by journalists as a 'risk story' or an official cover-up is suspected, then the authority of official sources may be undermined.

The above factors interact so that voices which gain access to the media at one point in time, may later be muted. For example, in 1990 the media were receptive to Professor Richard Lacey who appeared on television news and was widely quoted in the press arguing that BSE was dangerous to humans. Although failing to conform to some journalists' standard definitions of credibility, Lacey gained attention partly because of an official information vacuum. Journalists viewed MAFF as uncooperative and exercising unacceptable levels of control over the news agenda. After 1990, however, a political resolution with Europe and the absence of fresh scientific evidence, combined with consistent government reassurances, to mute coverage. Alternative voices were increasingly isolated and became 'old news'.

The relationship between expert/policy assessments and the media is two way; the media do not simply represent such assessments, they actually feed into assessment processes.

4.22 Media Influence on Research, Funding and Policy Priorities

Experts often commented that there were other equally pressing areas that needed to be given attention but that the media influenced what was given priority. Media attention may affect the research questions that are asked and the perceived value of research among peers, decision makers and funding agencies. It may also influence what is published in specialist journals. As an editor of one key journal commented, the mass media "*...stake out certain areas which we recognise, we position ourselves in relation to those areas*". Close involvement with the media can even alter the research which is carried out. One BSE researcher involved in a *World in Action* programme demonstrating that infected cows were still entering the food chain, commented that preparations for this programme included carrying out a number of tests which would not otherwise have been done (Dealler 1996).

4.23 The Media as a Risk Factor

Finally, the media are often seen as one of the risks to be assessed and managed. Both experts and policy makers alluded to the 'risks' media coverage posed to the economy so that policy making and presentation may be made with one eye on such potential threats. There is a further element within this process which needs to be addressed. That is the impact of 'public opinion' on policy assessments and on decision making and implementation. A number of points should be highlighted here.

4.24 Lobbying

Lobbying has forced certain risks onto the political agenda. It was not until June 1986, 7 months after the first diagnosis, that MAFF informed Ministers of the new outbreak and a further 10 months elapsed before the government moved to have the threat assessed. MAFF also attempted to keep the nature of the disease to itself for as long as possible. When MAFF announced the existence of the new disease in October 1987 it did so in the Short Communications section of the Veterinary Record (journal of the BVA). The government also kept tight control of information on BSE and journalists were given very little information. The effects of this policy were two-fold:

- secrecy and conspiracy became the key major news themes. Lack of information from official sources has made it easier for journalists to write about cover-ups and government inaction.
- it created a news vacuum which could be filled up by other sources, opening the way for different players to actively engage with the media to establish positions of credibility, to debate and ask questions.

Government action had the effect of mobilising people into action. So, for example, according to the BVA: *"There was pressure from everybody, there were pressures from veterinary surgeons, pressures from public health, pressure indeed from farmers in the South who were having multiple cases on farms, and I'm sure that there was almost panic within the department, what do we do now with this new disease?"*.

The very strictness of the official government line meant that those who 'disagreed' with it had to find ways of communicating their ideas. One scientist working on encephalopathies at the time said that:

> *What came from all this was that the media were the most efficient and effective way of getting anything done about a serious and dangerous disease. Mr. Waldegrave did not reply to letters, other MP's could not understand, government organisations (PHLS) had been told to do nothing…there was a consensus of ignorance among the medical profession and large numbers of experts who deliberately did not say anything, even though they knew the risk was bad…So, the media were the only route by which information could reach the population.*

So, willing alternative experts could easily be found and used as a balance to what little official information was being offered . This allowed the debate to widen and introduced a conflict at the level of science over the behaviour of the BSE agent and its potential consequences for animal and human life.

4.25 The Public as Voters/Consumers

Politicians may be most likely to become actively involved in promoting policy when the issue is seen as of widespread public concern. They are also increasingly seen as both an audience and a problem for expert and policy makers. A Health and Safety Executive report, 'Use of Risk Assessments Within Government Departments', describes how communicating to the public is viewed:

...pressure groups and the media have tended to exaggerate insignificant risks, often at the expense of ignoring larger more common-place ones. Departments agree that, as a result, the public is confused and treats government assurances of the safety...with considerable scepticism and agree that a government-wide strategy on risk communication would be beneficial.(Health and Safety Executive 1996, p.37)

Crucially, 'the public' are brought into risk assessment and policy making as consumers of goods and services. Such manifestations of 'public opinion' are combined with particular images of lay people and models about their relationship with experts/policy-makers. One dominant model presents lay people as irrational and ignorant, another views them as consumers of scientific information, a third emphasises the citizen's 'right to know'.

The perception of lay people as consumers of science, to some extent, as having 'rights' to information, is now influencing some policy. This is clearly seen to be of legitimate 'public interest' in a way that animal feeding practices were not. By contrast, perceptions of the public as irrational but powerful consumers have heavily influenced government responses to BSE. 'Public opinion', then, is rarely directly canvassed, but is the spectre at the table of many decision-making, implementation and publicity processes. How these processes operate depends on the model of public rights/expert duties adopted, the public opinion being assumed, the 'fit' between the risk-source and the specific individuals 'at risk', who is assigned responsibility, the resources and strategies of pressure groups and the potential effect of public reactions (e.g. whether they are direct purchasers of a 'risky' product).

4.26 Conclusion

The mass media are intertwined in the risk management process at every level. Experts and policy makers will emphasise media 'sensationalism', 'confusion' and 'obsession with risk'. However, the mass media are not automatically attracted to risk. Indeed, news media are ill-designed to give sustained attention to what Adams (1995) calls 'virtual risks'. Risks are unlikely to be newsworthy if they a) involve considerable uncertainty, b) are not subject to current policy reaction and c) may only be realised in the distant future. Assumptions that the media always favour 'extreme' or 'maverick' risk assessments are also incorrect. Different parts of the media will adopt different positions under diverse conditions, depending on varying newspaper or programme agenda, and all such representation will be provisional, not inevitable. Key factors work within the media which predispose them to give attention to certain types of risk assessments, and these can usefully be considered by policy makers and experts and by media personnel considering their own practice. However, focusing on the media as the problem risks losing sight of the relationship between the media and social institutions. The media are dependent on their sources. The shape and scope of issues in the press and on television are not just the result of media factors but also of struggles in the assessment and policy making processes themselves.

There is a need to reconnect media studies with the social context of media production. It is important to do this in ways which take into account changes over time. This temporal dimension is crucial both because of spiralling media momentum and eventual news fatigue and because of the dynamic interaction between events and media priorities. Understanding these processes allows some attempt at prediction about how future events might change coverage. Research needs to go beyond exploring 'discourses' or 'routines' to look at structures, events and practices which facilitate or inhibit the promotion of particular representations. Detailed documentation of these processes helps to account for both commonalties and variations in media reporting and points toward some limitations in general models of media production processes. In particular, standard accounts of journalists' hierarchies of credibility have limited explanatory value when accounting for risk reporting when, for example, lay or alternative voices can, on occasions, gain supremacy and official voices lose authority.

The literature on policy processes tends to focus on interactions between government and interest groups, neglecting the media's role. Similarly, sociological theories about risk have given only cursory attention to relationships between experts, policy makers and the media. Indeed some grand theory about, for example, a "generalised climate of risk" (Giddens 1991), obscures discussion of why some risks appear more significant than others as well as why risks emerge, and fade away, when they do. Empirical examination can challenge some generalisations and illustrate the ways in which the media are now crucial to the political process and operate as a space in which struggle over the discovery and definition of risk can be fought out.

This is one way of considering how risk is perceived in different ways in the media and how these perceptions change, partly as they conflict with one another. But, however these perceptions were organised and portrayed there was an underlying reality – the disease of BSE and the real deaths of real cattle. While there were scientific questions about the nature and cause of the disease and its ramifications as regards vertical and cross-species transmission, which had its own continuing uncertainties, the evidence the growth of the disease beyond official predictions impinged on perceptions of what was going on, not least on consumers. As we think about these processes we can see that it is one way of trying to understand the inter-relation between the objective and subjective aspects of social reality. It does give us some understanding of the processes of social amplification in this case.

Acknowledgements

The author would like to acknowledge the collaboration of Jenny Kitzinger, Reader in Media Studies. Brunel University, on some of the research reported in this chapter.

References

Adams J (1995) Risk. UCL Press, London

Adams W (1992) The role of media relations in risk communication. Public Relations Quarterly, Winter 1992–1993: 28–32

Banks J and Tankel J (1990) Science as Fiction: technology in prime time television. Critical Studies in Mass Communication 24(36): 24–36

Beck U (1995) Ecological Politics in an Age of Risk. Polity Press, London

Dealler S (1996) Lethal Legacy: BSE – the search for the truth. Bloomsbury, London

Dunwoody S, Peters HP (1992) Mass media coverage of technological and environmental risks: a survey of research in the United States and Germany. Public Understandings of Science 1: 199–230

Friedman S, Dunwoody S, Rogers C (1986) Scientists and Journalists: reporting science as news. Free Press, New York

Giddens A (1991) Modernity and Self Identity. Polity Press, London

Goodell R (1987) The role of mass media in scientific controversy. In: Engelhardt T, Caplan A (eds) Scientific controversies: case studies in the resolution and closure of disputes in science and technology. Cambridge University Press, Cambridge

Hansen A (1994) Journalistic practices and science reporting in the British press. Public Understanding of Science 3: 111–134

Health and Safety Executive (1996) Use of Risk Assessments in Government Departments. HMSO: 37

Jacobs R (1996) Producing the news, producing the crisis: narrativity, television and news work. Media, Culture and Society 18 (3)

Kitzinger J, Reilly J (1991) The rise and fall of risk reporting: media coverage of human genetic research, 'false memory syndrome' and 'mad cow disease'. European Journal of Communication, 12 (3)

Lichtenberg J, Maclean D (1991) The role of the media in risk communication. In: Kasperson R, Stallne P (eds) Communicating Risks to the Public: international perspectives. Kluwer Academic publishers, London

Miller D, Reilly J (1995) Making an issue of food safety: the media, pressure groups and the public sphere. In: Maurer D, Sobal J (eds) Eating Agendas: food and nutrition as social problems. Aldine de Gruyter, New York

Peters HP (1995) The interaction of journalists and scientific experts: co-operation and conflict between two professional cultures. Media, Culture and Society 17: 31–48

Rogers E, Chang SB (1991) Media coverage of technological issues. In: Wilkins and Patterson op. cit.

Sandman P (1988) Telling Reporters About Risk. Civil Engineering 58: 36–38

Schanne M, Meier W (1992) Media Coverage of risk: results from content analysis

Schoenfeld A, Meier R, Griffin R (1990) Constructing a social problem: the press and the environment, Social Problems 27: 38–61

Schlesinger P (1990) Rethinking the sociology of journalism: source strategies and the limits of media centrism. In: Fergurson M (ed) Public Communication: the new imperatives. Sage, London

Southwood R (1989) Report of the working party on Bovine Spongiform Encephalopathy. HMSO, London

Stallings R (1990) Media Discourse and the Social Construction of Risk. Social Problems 37: 80-95

Wilkins L, Patterson P (1991) Risky Business: communicating issues of science, risk and public policy. Greenwood press, London

Exploring Attitudes to Eating Fruit and Vegetables

5

TERESA BELTON

5.1 Introduction

"An apple a day keeps the doctor away", is a well-known adage; indeed, the health benefits of eating plenty of all kinds of fruit and vegetables appear to be legion and have been widely disseminated. So it is puzzling that fruit and vegetable consumption for most people in the UK, as well as much of Europe and the US, is well below recommended levels. When addressing the gap between expert advice and individual behaviour, this issue is therefore one that plainly deserves examination. This chapter explores the reasons for the clear discrepancy as examples of the kinds of subtle and complex factors which can get in the way of people acting on apparently sound nutritional information handed down by scientists.

There is now a very substantial body of evidence to suggest that eating sufficient fruit and vegetables significantly reduces the risk of developing coronary heart disease and a variety of cancers, the two most common causes of death in Britain (World Cancer Research Fund 1997, Committee on Medical Aspects of Food and Nutrition Policy 1998). Several studies have shown that diets rich in fruit and particularly, perhaps, vegetables, are associated with a reduced risk of a variety of tumours, especially cancers of the respiratory and gastrointestinal tract (Lambert 2001). Micro nutrients found in these plant foods are also important for healthy foetal development, both before and after conception (Dallison and Lobstein 1995). The minimum daily intake of fresh fruit and vegetables (including fruit juice but excluding potatoes) which will have a protective effect against cancer and coronary heart disease has been calculated at 400 g (World Cancer Research Fund and American Institute for Cancer Research 1997), and since the mid 1990s health promotion campaigns in the UK, US and some European countries have been publicising this knowledge and attempting to turn it into a form that is easily applied in practice by encouraging people to eat at least five helpings of fruit and vegetables every day.

Following such advice might seem, at first sight, to be an easy means for individuals to exert a major beneficial influence on their own and their families' health. But statistics show that the majority still fall well short of the recommended daily intake of these simple foods (Ministry of Agriculture Fisheries and Food 1999). This chapter makes a wide-ranging exploration of reasons why people in Britain (and elsewhere in the Western world) often do not act fully on such advice from health experts. It divides into three parts; the first part considers the practical barriers that can hinder the eating of plenty of fruit and vegetables and takes a critical look at a wide variety of other factors which shape personal and cultural attitudes to food and health and which can constrain consumption of fruit and vegetables. The second part reports a recent study which provides some empirical examples to illuminate relevant issues; and the third part suggests that a radical change of political, social and culinary culture is required to bring about a significant general increase in fruit and vegetable consumption.

5.2 Factors Recognised as Influencing Fruit and Vegetable Consumption

5.2.1 Poverty

5.2.1.1 *Low Income*

The National Food Survey (Ministry of Agriculture, Fisheries and Food 1999) shows that levels of household expenditure on fresh fruit and vegetables reflect household income – generally speaking, the better-off a household is the more it spends on fruit and vegetables, and vice versa. This would suggest that income is a critical factor in determining the amounts of fruit and vegetables that people eat. Indeed, organisations concerned with poverty have made a strong case for addressing poverty on the grounds that it affects basic nutrition. In recent years poverty has been experienced by a growing proportion of the British population; a report published by the Joseph Rowntree Foundation (Gordon et al. 2000) found, using government low income data, that, by the end of 1999, 26% of the British population were living in poverty, (in comparison with 14% in 1983). Poverty was measured in this study not only in terms of income but also of deprivation of three or more necessities, defined as items deemed essential by more than 50% of the population; these included such things as a bed and bedding, two pairs of shoes, a telephone and being able to visit family and friends in hospital and attend weddings and funerals.

The premature death rate (death before the age of 75) from coronary heart disease, is 58% higher among male manual workers, such as builders and cleaners, than among male non-manual workers such as doctors and lawyers, while among women the death rate is double for manual workers compared with their non-manual sisters (British Heart Foundation 1999). The picture for cancer is similar (Department of Health 1998). It is the disparity in **income** between manual workers and professional workers which is the clearest difference between these two groups.

Poverty can be experienced in many ways, such as poor housing, an unclean and unsafe environment, unemployment, and lack of hope, as well as a poor diet, and the reasons for the higher morbidity and mortality rates amongst the less well off are complex. However, there is a strong consensus that diet is a critical factor in determining an individual's long-term health. The ability to afford a healthy diet is of particular importance for pregnant women, as inadequate nutrition during gestation is linked to an increased risk of low birth weight, itself the likeliest cause of perinatal death (Dallison and Lobstein 1995). Low birth-weight babies who do survive suffer from higher than average rates of childhood illness and disabling conditions such as visual impairment, deafness, cerebral palsy, behavioural disorders and mental retardation. According to research carried out into poverty and undernourishment in pregnancy for the Maternity Alliance and NCH Action for Children (Dallison and Lobstein 1995),

> *[T]he consistent theme to emerge from all the research reports is that poor diet is linked to low income. Social class, regional variations, dietary, religious and cultural preferences, and educational achievement have a limited or nominal impact in comparison ... [It] is virtually impossible for a woman to adequately feed herself when on Income Support (pp. 12–13).*

The result of low income in dietary terms, according to the analysis in this report is that, rather than the minimum 5 helpings of fruit and vegetables a day recommended by the Department of Health, the average intake of fruit and vegetables among these women was 4 pieces of fruit, 3.5 helpings of green vegetables and 1.5 servings of salad **per week**. Moreover, 94% of the women in the study said they missed at least one whole meal a week due to lack of money. Most spent less than £ 2.00 a day per family member on food, and a large number less than £ 1.50, while the national average at the time was just over £ 3.00. More generally, benefit levels for all groups are set without any calculations which relate them to the costs of a healthy diet (Webster 1998). It has been suggested that if money is short, one immediate way of reducing expenditure is to buy cheaper food, and less of it (Dallison and Lobstein 1995). Foods that supply the most calories (i.e., the most energy or the most satisfied feeling in the stomach), at the least cost, are fatty and/or sugary foods such as cakes, biscuits, pastries, snack foods, confectionery and soft drinks. Plain lack of money is clearly one reason for eating inadequate amounts of fruit and vegetables.

5.2.1.2
Inadequate Access to Supplies of Fruit and Vegetables

The simple cost of food, however, is not the only relevant factor when relating the eating of fruit and vegetables to income. 'Food poverty' also encompasses another practical barrier to acquiring healthy food, namely physical access. Since the 1970s the trend for supermarkets to take over from high street shops and markets has constituted a marked social change. Supermarkets have increasingly been built on out-of-town sites, robbing communities of most of their small, specialised shops, which have shut down in large numbers, unable to compete with the convenience and economies of scale on which supermarket trade depends. Thus, the number of local greengrocers has declined drastically (from 26600 in 1971 to 12700 in 1992) This trend has created 'food deserts' in residential areas of towns and cities and left the majority of rural communities without a village shop (Lobstein 1997). For people unable to afford a car, shopping in a supermarket is likely to require at least the journey home to be taken by bus or taxi. Taxi fares add significantly to the cost of shopping on a tight budget, and bus rides, which can be unreliable in the current state of public transport, can add considerable physical demands to a shopping trip. It has been found that, whereas for social class A and B shoppers, easy parking is the paramount factor in determining which shops are patronised, for social class E shopping habits are determined by low prices, followed by closeness of shops to home (Lobstein 1997). Buying enough vegetables to provide all family members with the minimum recommended quantities might seem an impossible demand when having to carry them on a bus in addition to other bulky basics such as milk, potatoes, bread, cleaning materials, washing powder and toilet paper, as well, perhaps, as a toddler and a pushchair. Under these circumstances, much in the way of fresh fruit and vegetables may well seem dispensable. Those corner shops that remain have had to raise their prices to try and offset the loss of custom to supermarkets, and the selection of fruit and vegetables they offer is usually lacking in both range and freshness, and is thus often quite unappetising. Meanwhile, supermarkets have expanded their ranges to include expensive, exotic commodities new to this country, and pre-washed, pre-chopped packs of salad, stir-fry and other vegetables and fruits for the cash-rich, time-poor shopper. Quantities of organic produce are also rapidly increasing for those with the financial resources to exercise new choices.

Poverty clearly makes the purchasing of adequate quantities of fruit and vegetables an impossibility for a significant proportion of the population. Policies which reverse the growing inequality in income in Britain are therefore urgently needed to end the situation in which some people simply cannot afford basic nutrition. For people without the use of a car or good public transport, and those in rural areas, lack of access to supplies of good quality fruit and vegetables at reasonable prices is a major disadvantage.

It seems, then, both from the evidence of research conducted among people living in poverty and from the circumstantial evidence of current shopping conditions that low income is the overriding factor which stops people from consuming adequate amounts of fruit and vegetables. There are, however, also signs that lack of money, or access, is not the whole story. It is interesting to note that, while 55% of the women on Income Support who participated in the research on poverty and pregnancy by Dallison and Lobstein cited above claimed that the reason for their not eating healthier foods was that they were too expensive, another 46% said the reason was that they did not like them, and only 6% said that it was because they were difficult to buy. Indeed, contrary to the notion that poverty precludes the consumption of healthy food, the very literature that testifies to the actual experience of poverty offers evidence of a determination of some mothers to feed their families healthy quantities of vegetables, despite highly circumscribed means. *Mother Courage: Letters from Mothers in Poverty at the End of the Century* (Gowdridge et al. 1997) is a collection of letters recording the experience of women who are, for a variety of reasons, struggling financially while raising their children. One writes,

We eat okay but very simply, meat is an occasional treat, we eat lots of pulses and pasta, tons of fruit and vegetables. I'm obsessive about getting enough fresh fruit and vegetables inside us … We rarely go out and I've lost some friends through this … and I hardly ever get new clothes (p. 33).

Another, a single mother of three, exclaims,

Quite honestly it makes me very mad to watch these programmes on TV that harp on about how poor every single parent on benefits … is … [Single parents] moan they cannot feed their children – do they expect to feed them steak or/and joints of best beef? … Well I agree the benefit barely covers these [necessities], but with good management one can do it. Albeit there is nothing to spare for [the] pleasures … But I don't expect it. Why should the government pay for these things when they had no hand in me getting pregnant! … And we have everything we need and me and the children feed well and are clothed well (p. 182).

The words of these women seem to express a tenacious belief in the necessity of simple, healthy food for their children above all else. But such values and determination seem to be the province of a minority, and most people facing difficulties in feeding their families need support.

5.2.1.3
Positive Developments With Regard to 'Food Poverty'

Some constructive initiatives to support those in particular need have, in fact, been stimulated by negative social and economic changes. Three forms of action, which tackle food poverty from various perspectives, are outlined below.

One type of initiative has been the development over recent years of 'food projects', a variety of enterprises operating in particular communities, sometimes involving professionals, which have helped some people on low incomes to access and enjoy a better diet. The means employed cover a range of activities which include food co-ops for the bulk buying of fruit and vegetables at reduced prices, cook and eat sessions, community cafes, nutrition education among new mothers, and projects which grow food, and provide children's breakfasts and adult evening meals. The Joseph Rowntree Foundation commissioned a study of the factors which help or hinder food projects in becoming established and being maintained (McGlone et al. 1999). It found that 70% of users participated for social reasons: to meet new people, have fun or get out of the house; about half did so because the projects were convenient and sold food at low cost; only 20% did so in order to enable them to have a healthier diet. Although no formal evaluation of the impact of these projects has been carried out, it was informally found that most people felt positive about them and there was anecdotal evidence of some degree of effectiveness. Users of the cook and eat sessions said they gained confidence and new skills, trying new recipes as a result; food co-ops made fruit and vegetables more cheaply and easily available so that users bought more and felt able to experiment within a budget; and all those interviewed reported social and psychological benefits. The report concludes that,

> *The social gains at individual and community levels are not separate from nutritional outcomes but intrinsic to their achievement. Overcoming social isolation, giving people a sense of worth and well-being, empowering them, and raising levels of skills and training enable individuals to feel in more control of their own health and welfare. There is then the possibility to implement changes and move towards healthier eating* (p. 41).

While localised food projects have proved their potential to help small groups of people move towards healthier diets, a more universal approach is required for there to be a real impact on that large section of the population, estimated at 4 million people, who are currently denied the chance to feed themselves adequately. This situation has begun to be addressed by the Food Justice Campaign, a coalition of concerned groups, launched in November 2001. The campaign, which includes members of parliament, is sponsoring a 'Food

Poverty (Eradication) Bill', the aim of which is to make it a legal duty for the government to make fresh food more available and more affordable, by ensuring a co-ordinated strategy to support local sale of local produce, improve transport and reverse the decline of small shops.

Another development, propelled by the need to ensure both sustainable agriculture and a sustainable environment, has been the growing recognition of the value of selling and buying food close to its source of production. The new trend is powerful: farmers' markets, first set up in 1997 specifically to enable producers to sell their own farmed, grown or produced food direct to the public, numbered 16 by February 1999 and increased to over 200 less than a year later (National Association of Farmers Markets).

5.2.2 Health Education

In order that the lay public should be encouraged to buy and eat fruit and vegetables for the sake of their health, they must, of course, receive information about the nutritional importance of these foods. The next section will therefore look at the role of health education and why the relationship between knowledge and behaviour is not necessarily straightforward.

5.2.2.1 *The Valuing of Health and a Healthy Diet*

Among those who unequivocally place a high value on their own and their families' health, especially those with a professional interest in any aspect of health promotion or activities which have implications for public health, it may be taken for granted that good health is important for others also. But, as the large-scale Health and Lifestyles survey shows (Blaxter 1990), not everyone values health equally. Interviews with 9003 individuals revealed that the relationship between attitudes, beliefs, behaviour, age and other relevant social and economic factors is, in fact, extremely complex. About two-thirds of interviewees in that survey were of the opinion that people are responsible for their own health. According to this view it is individual behaviour that determines state of health and development of disease, rather than external factors, such as air quality, over which one has no control. However, this attitude was less strongly expressed when questioning moved from the abstract to respondents' own health. Those who thought that people were responsible for their own health were, not surprisingly, more likely to have good diets. For the population as a whole, however, according to Steptoe et al. (1995), health considera-

tions are certainly not more important in determining food choices than other factors, such as sensory appeal, price and convenience.
Blaxter's Health and Lifestyles study (1990) found that there were two groups among whom a lack of positive attitudes towards personal health and unhealthy behaviour could be associated. The first group were mainly young men, who did not appear to think about health at all and who were likely to think their health "good" even though it was often less than perfect. They might well be aware of public health issues, but did not see their relevance to their own lives. This is in keeping with the finding of Walker (1997) amongst the adolescent participants in a study of boys' sexual health education and attitude change. This group of young males generally put much emphasis on "fitness", which they saw as something to be proud of and which they believed was in their control, as opposed to illness which they saw as a weakness and beyond their control. Thus, although they knew that "junk food" and smoking were bad for their health, they felt little motivation to change their habits. The second group in the Health and Lifestyles survey with negative attitudes were people in poor health. Likely to be women over 60, living on low incomes, and in troubled social circumstances, this group believed that health was important and, in general, a consequence of behaviour, yet they had very unhealthy habits and believed that their own health was beyond their control.

5.2.2.2
The Relationship Between Health Education and Behaviour

Although an individual's general level of formal education has been shown to have a significant bearing on their diet, those with higher qualifications tending to eat more fruit and vegetables, evidence of the impact of information and advice specifically intended to improve eating habits is not at all clear. Any attempt to increase public knowledge with the aim of improving public health is based on the implicit assumption that enhanced knowledge will lead to changes in behaviour. But, according to Kuhl and Beckmann (1985),

> *The link between cognition and behaviour is less reliable than one might expect. People do not always perform in a manner consistent with their beliefs, values, attitudes or intentions … many investigators report disappointingly low attitude-behaviour correlations across a great variety of situations. Cognitive models of motivation that explain an individual's behaviour on the basis of her or his expectancies and values do not have high predictive power* (p. 1).

This view seems to be borne out by a meta-analysis of 9 studies attempting to relate knowledge to attitudes and dietary intake (Axelson et al. 1985) which found statistically significant but only small associations between attitudes and behaviour. The correlation between knowledge and behaviour was even lower. A clear example of how belief is no predictor of either attitude or behaviour is the knowledge of virtually all smokers that smoking is very likely to damage their health. In the case of fruit and vegetables, the extensive US National Five-a-Day Program seemed to make significant improvements in the public's awareness of the link between fruit and vegetable intake and disease prevention and the shortcomings of their own diet, but did not appear significantly to increase actual consumption of these foods (Gibson et al. 1998).

Responses to the Health and Lifestyles survey undertaken by Blaxter led her to conclude that, "It seemed that the public had learned well the lessons of health education, and answers about 'healthy lifestyles' were the ones which come first to their minds, or ones which they saw as the 'correct' or expected replies" (Blaxter p. 153). Similar conclusions emerge from the study of pregnant women by Dallison and Lobstein (1995) who comment, "Although their understanding of what constitutes a healthy diet was incomplete, ... it was clear that most women did understand the importance of vegetables and fruit" (p. 29). In general it seems that almost everyone realises that fruit and vegetables are "good for you", but there is much less understanding of the specific benefits which they provide or the degree to which they do so (e.g. Cox et al. 1997; Brug et al. 1995b), or motivation actively to incorporate that knowledge into daily eating decisions.

A recent study (Lawton in preparation 2001) illustrates how knowledge of what constitutes a healthy lifestyle is not sufficient to ensure that sound advice is followed. The study, in which adults with a wide spread of ages and occupational backgrounds were interviewed about their perceptions of illness and death and the effects of these on their health-related behaviours, found, in common with other research, that people generally take good health for granted until something happens to shake their confidence in its continuance. There is a natural tendency for older people to be more aware of the likelihood of illness and the inevitability of death but, more than advanced age, the study found that actual bodily experience of acute or chronic illness is the factor most likely to bring home to people their vulnerability to disease. Although participants were well versed in disease prevention messages, their adoption of health promoting lifestyle changes, including the eating of more fruit and vegetables, was, with almost no exceptions, reactive rather than proactive, brought about as a result of actual ill health. "It became apparent that long-term health considerations were not of central, if any, salience to these participants" and "their comments often indicated a perception that health promotion activities lacked any direct and immediate relevance to themselves." Calnan and Williams (1991), too, have found that health-related dietary changes most often take place in response to actual health problems such as heart disease. They did find, however, that middle class respondents were more likely to have made changes to health-related behaviours in response to knowledge of the **potential** dangers to health of not making those changes than were working class respondents.

5.2.2.3
Inconsistency Between Knowledge and Perception

Even with an awareness of the importance of a healthy diet in general and of fruit and vegetables in particular, many people have repeatedly been found both to be uncertain as to the daily quantities of fruit and vegetables actually recommended and also to overestimate their own consumption. Indeed, independent of the actual amounts they were eating, almost all the 29 participants in a study carried out in the Netherlands considered themselves to be eating enough (Brug et al. 1995a). A similar picture emerged from a questionnaire survey of 2020 UK adults intended to be representative of the whole population (Cox et al. 1998). This found that, regardless of current intake, the majority of respondents saw themselves as consuming more fruit and vegetables than the average person; in fact many individuals in the lower intake categories perceived themselves as having high intakes. A study by Gibson et al. (1998) which measured the actual fruit and vegetable consumption of mothers and their children aged 9–11, illustrates the complexity of the relationship between knowledge and eating behaviour. These participants overestimated their consumption by about 30%. While mothers' and children's nutritional knowledge was positively related, and mothers' nutritional knowledge was positively and quite strongly related to children's fruit consumption, it was not related to their children's eating of vegetables. In fact, mothers' and children's ratings of healthiness did not correlate with children's intakes of respective food types. The only exception was that, while children who considered health to be important for the foods they chose ate less confectionery than others, those who gave vegetables the highest healthiness ratings tended to eat the **least** vegetables. Such research suggests that being informed about the health benefits of fruit and vegetables does have some impact on consumption for some people, but that human beings cannot be expected to be entirely objective, accurate or uniform in their judgement of food consumed. Indeed, they do not necessarily act on a purely reasoned basis, personal taste weighing at least as heavily as rational motivators in determining food choices.

5.2.2.4
Why Advice on Diet can Never Hope to Reach Everyone

Whatever their reasons, it seems that there are some people, as can be seen from the evidence of the Health and Lifestyle (Blaxter 1990) survey, who do not regard cultivating good health as particularly important. No amount of information aimed at them by well-meaning authorities is likely to make any impact on them, for, regardless of the intention of the transmitter of that information, each receiver of the message will receive it in their own way. In the early days of mass communications it was naively assumed that a media communication was a 'magic bullet' that carried a single message that would have a universal effect on all audience members. Now, however, there is a more sophisticated understanding that accepts that a number of active selection processes are undertaken on the part of the recipient of media communications (Severin and Tankard 1992). The first is selective exposure, the tendency to seek out messages that support one's own view and avoid those which contradict it. Selective attention reinforces this. Then comes selective perception, or interpretation. Finally, selective retention ensures that we remember more readily that information which accords most with our wants, needs, attitudes and assumptions. On this basis, those people who particularly value health and believe that its maintenance is their own responsibility will pay more attention to nutritional advice, while those who live their lives according to other priorities will not notice or absorb efforts to impart health education. An experiment in the US suggested that what determines this difference in response to dietary or health advice is not so much socio-economic status or educational level as the degree of pre-existing interest in a particular topic – due, for instance, to suffering from a particular condition (Severin and Tankard 1992).

Several discrete types of behaviour have a direct influence on health, principally smoking, exercise, heavy consumption of alcohol and eating habits. People weight these differently. The Health and Lifestyles survey (Blaxter 1990), conducted in the late 1980s, showed that at that time, at any rate, smoking and exercise were given more importance than diet and drinking. People are therefore likely to take more notice of messages about some behaviours than others. However, these messages still do not always have the desired effect, for the Health and Lifestyles survey found that those with the relevant unhealthy behaviour were very conscious of its link with disease. Sometimes immediate rewards can seem to outweigh long-term costs, and unhealthy behaviours may play an important part in individuals' coping strategies. Indeed, Blaxter concluded that the potential of health education is limited because social circumstances appear to be a major factor in quality of health. However, this is not a reason to give up the dissemination of information and advice but rather grounds to keep trying to improve social conditions. After all, 47 % of respondents to the Health and Lifestyles survey claimed to have made some changes to their health-related behaviour in the past 10 years had taken significant steps to alter their eating habits in healthy ways. This tendency is presumably not a matter of chance but a direct or indirect result of health promotion activities which had enhanced their awareness of health issues.

5.2.3 Food Culture

While health education may exert an overt influence on people's diets, a much more subtle, but nevertheless powerful, influence on attitudes to food and eating results from the prevailing culture. The dominant culture at any one place or time derives from much broader and deeper-seated factors than the current state of knowledge on nutrition. Some of these factors are now considered.

5.2.3.1 *Between and Within Societies*

Patterns of fruit and vegetable consumption vary considerably across the world and also within countries. A meta-analysis of studies aimed at elucidating disparities in consumption across Europe found that, as in Britain, consumption was generally higher among the better educated. In southern and eastern Europe, however, the reverse was true in some areas, where overall fruit and vegetable consumption is much higher than in the rest of Europe (Roos et al. 2000). The reason for the inversion of the relationship between education and fruit and vegetables in the diet in these areas could be that those in the lower socio-economic classes may have better access to cheaper produce. Or, because these foods are so common, they may lack status and therefore lack appeal for the socially ambitious.

Research into the diets of refugees and asylum-seekers recently arrived in East London from Somalia, Kosovo, Colombia and Ecuador (Sellen et al. 2000) has also clearly shown that culture is a major influence on the amount of fresh fruit and vegetables included in the everyday diet. The study found that the diets of the refugees were more in line with current nutritional advice than those of the local population on similar incomes, relying on fresh foods and making little use of pre-packaged and cooked food outlets. Tasty and varied meals were achieved by tight budgeting, sound cooking skills and help from family and friends. Interestingly, though nearly half the mothers questioned thought that milk was the most important food for young children, 23% prioritised fruit and vegetables. Only half of them associated diet with health, however, which suggests a weaker connection with education and a stronger connection with tradition or culture than is generally concluded to be the case from interpretation of Western European data.

One of the reasons that the refugees were able to provide their families with plentiful fresh food was that they were confident of their cooking skills. In contrast, lack of these basic skills has been given as a reason by some western people for their low consumption of vegetables (e.g. Brug et al. 1995a). A critical erosion of the level of cooking skills among a significant proportion of young adults is one cultural characteristic which has changed in the west in recent decades, no doubt influenced by the interrelated factors of women increasingly leaving the home for the workplace, greater affluence and the orientation towards 'convenience' foods. Long gone is the time when the need to eat or to provide food required that all the processing of raw ingredients into edible fare be done in the home.

Even if a lack of culinary skills is not a factor for everyone, those living or eating alone or with little time available for cooking – again increasing trends in the West – are less likely to want to go to the trouble of a great deal of vegetable preparation. A focus group study from the Netherlands (Brug et al. 1995b) discovered that food-providing behaviour can be strongly influenced by other people, so that mothers do not buy fruits and vegetables that their children or partners do not like, while people offering hospitality are likely to go to more than normal lengths to provide their guests with a good array of cooked vegetables or salad. It also found that some people see fruit as awkward or messy rather than an easy snack. People often consider fruit not to be filling, and there is a perception among some that 'vegetable dishes' and fruit are expensive (Cox et al. 1998). Habit is also important in human behaviour; indeed, Brug et al. (1995) found in focus group interviews which aimed to elicit the psychosocial determinants of fruit and vegetable consumption that,

> *It was said repeatedly during the interviews that they were eating the way they were taught at home in the past and continued eating according to those habits when they left their parents to go and live on their own or started their own family* (p. 104).

Those, moreover, who habitually include generous quantities of vegetables and fruit into their regular eating patters are likely to find the attendant demands on shopping and food preparation less onerous than those who do not have this habit and who therefore have to make a special effort to provide and eat more of these foods.

5.2.3.2
Public Food

What one might call public, everyday food culture in Britain is largely characterised by what is provided by public institutions, notably those devoted to health and education, and that offered by high street eating places. Neither has been a model of good nutrition in recent times, particularly with respect to vegetables and fruit. Burger restaurants and fish and chip shops, whose menus include little of either, are commonplace and very popular. And hospitals, where one would hope the therapeutic importance of diet would be understood, have become notorious for unappetising fare, including sloppy, overcooked food (Carlowe 2001). That the practice may not match the rhetoric was illustrated by the personal experience of a close relative of mine which showed that what was given to patients could actually run counter to the hospital's own advice: on a cardiology ward, while the door displayed a large poster exhorting patients to eat less dairy products, biscuits, cakes and sugar, and to eat wholemeal bread instead of white, he was presented with a plate of white bread cheese sandwiches, a packet of biscuits and a drink of squash; the saving grace was an apple. School meals, too, after nutritional standards were dropped as the service was required to become self-financing in the 1980s, began to provide chips, burgers and sausages as standard savoury dishes, and desserts have tended to include a large proportion of fatty, sugary items. School caterers have found that they have had to provide such options daily as these are what many children are used to eating and they refuse healthier alternatives, preferring what is familiar – as research shows is the pattern to be expected (Steptoe et al. 1995). Vegetables and fruit have therefore not been prominent amongst the foods provided by school meals.

There are now, however, moves underway that do begin to address public nutritional needs in a positive and proactive way. In 2000 the UK government set up the Better Hospital Food initiative to overhaul the catering service provided for inpatients. This recognised both the extent of food wastage in hospitals and changing patterns of eating and lifestyles. A National Health Service website (http://www.betterhospitalfood.com) explains that the reforms now being introduced have been developed by the National Health Service Menu Group, which includes caterers, dieticians, nurses and some leading chefs, and are based on extensive research and menu testing among all kinds of patients in many hospitals. This has included the interviewing of 612 patients in 3 hospitals, of whom 31% said they wanted more choice of fresh fruit for breakfast, 23% wanted a vegetarian dish for the main meal, 63% said they would enjoy eating salad (compared, e.g., with 44% who would enjoy fish fingers), and 48%, the highest percentage, said their preferred pudding was fresh fruit. Fruit was the second favourite snack, after sandwiches and ahead of crisps, cake, pastries and chocolate bars. These results appear, to some extent, to contradict preferences revealed by actual consumption figures (e.g. Ministry of Agriculture Fisheries and Food annual National Food Surveys) and research reported here, thus raising questions about the sample of patients interviewed and whether the hospital context of the survey affected responses in some way.

However, if the new provision really does take patients' reported wishes into account then salads and fruit should come to feature more prominently in hospital food than previously.

The need for a somewhat better informed approach to school meals provision has also been recognised, and the government did introduce compulsory minimum nutritional standards for school lunches on 1st April 2001, which require the provision of at least one vegetable or salad every day and fruit-based desserts at least twice a week. Thus children will have the opportunity to eat somewhat more healthily at school, if they choose to do so.

5.2.3.3
Advertising

One factor in purchasing and consuming culture in contemporary Britain and other western countries is the highly influential role of advertising. However, fruit and vegetables are rarely the subject of advertising campaigns as they are not usually associated with brand names. Outspan oranges have been one of the few exceptions, but apples, tomatoes, cabbage or carrots are hardly likely to be specially promoted by retailers beyond the occasional "buy one get one free" offers on more expensive items, such as melons, in supermarkets, or bags of bananas for 50p in a market late on Saturday afternoon. Although advertising makes commodities more expensive it is very effective in selling them, such are its powers of persuasion. An inquiry by the Co-operative Wholesale Society (2000) found that, while fatty and sugary foods should make up a mere 7% of what we eat, 70% of all television advertisements during children's peak viewing times were for such foods, and of course there was none for fruit or vegetables. Indeed, on Saturday morning, when more food and drinks commercials were shown per hour than at any other time, 99% were for foods high in sugar, salt or fat. The enquiry also found that children responded positively to them; market research with 293 children under 11 years of age, and group discussions with adults and children revealed that 73% of the children asked a parent to buy the sweets or crisps advertised, and all but 3% showing some level of interest, and that if parents refused, 29% of children kept pestering and another 22% got upset or cried. Such behaviour puts parents under considerable pressure, which they are not always able to resist. Consuming these highly advertised products leaves both less room in the stomach and less money in the purse for healthier snacks such as bananas or raisins, and trains the palate to desire more of the same (Conner 1994). The continued advertising of other, more substantial processed dishes to adult consumers may well take up financial resources and diminish inclination to eat simpler foods that would require more effort and planning to prepare.

Advertising can take a more insidious form, too, implicitly purporting to offer a service: the Food Commission reported that, as children went back to school in September 2000, a leading supermarket chain took 3 pages of its in-house magazine to promote 17 food products for children's lunch boxes, including processed cheese, chocolate biscuits, flavoured drinks and crisps. 10 of these products contained very high levels of fat and saturated fat and others had elevated levels of sugar. Again, there was no promotion of fruit or vegetables.

5.2.3.4
Snacking

A recent development is the burgeoning phenomenon in the UK, of vending machines supplying fizzy drinks, crisps, chocolate bars and so on, in all sorts of public places, both indoors and out. These encourage snacking and the consumption of exactly those foods of which health experts advise us to eat less. They compete directly with fruit as snacks, and can blunt the appetite at mealtimes, or even encourage people to eat snacks in the place of some meals. Such a trend, being purely commercial, cannot, of course, be vetoed by health authorities, and would not develop without a positive response from consumers. Research in the US (Zizza et al. 2001), indeed, bears out that, as amongst children, there has been a large increase in both the number of snacks and the calorific value of snacks consumed by young adults in recent years. Consumption of such foods might in turn prime a desire for further consumption due to familiarity and aided by easy accessibility. Wide public availability of snack foods creates a frequent temptation, as easy to find in health and education establishments as elsewhere, and could implicitly reinforce a particular attitude towards eating or the notion of food. It could even be claimed that in certain affluent parts of the world two different types of consumption have developed and become confused, that is, eating and drinking for nourishment (which can be enjoyable as well as nutritious), and eating and drinking purely for sensory pleasure.

5.2.4
Personal Taste

What constitutes sensory pleasure is a matter of personal taste, and it must not be forgotten among considerations affecting large groups of people that individuals vary enormously amongst themselves in all matters of taste. Individual food preferences must play an important part in eating choices. It has been frequently reported that the overriding reason people give for choosing a particular food to eat is enjoyment (e.g. Shepherd and Sparks 1994; Rappoport et al.

1992). So pleasure almost always comes before health in the hierarchy of motivation in selecting food. Thus, conversely, the greatest barrier of all to eating plentiful vegetables and fruit is not actually liking them. How tastes for particular foods are developed is still incompletely understood, but research can offer some pointers. It seems that eating patterns are largely established in childhood. For example, a recent study found that elderly Greek Australians have tended to retain several features of the Mediterranean diet with which they grew up (Lairon 1999). Besides the external culture of childhood, the particular circumstances and mores of the individual home will determine the nature of a child's encounters with food, and these, too, can exert an influence on long-term eating patterns. One study found, for instance, that working mothers stop breastfeeding earlier and rely more on commercial baby foods than do non-working mothers, and such experience, the authors suggest, could influence food choices among pre-school children (Earland and Ibrahim 1995). By the same token, it seems reasonable to propose that the increase in the number of households in which the parents or parent work full-time might well have entailed a reduction in the regular cooking of fresh food, which, in turn, might affect the eating experience and tastes of children. Another study compared the food rules among 838 women respondents to a questionnaire, leading the researchers (Hupkens et al. 1995) to conclude that upper-middle class mothers are likely to impose stricter rules in relation to eating healthily than lower class mothers.

That simple exposure of young children to particular foods affects their response to them, has been experimentally demonstrated. A study was carried out by Fleming et al. (1995) who repeatedly presented twelve previously rejected fruits and vegetables over a number of days to 5–7 year-olds during evening meals at home. The children were subsequently shown videos in which older children were seen eating them and encouraging the viewers to do likewise. The younger children were given rewards if they did so. The result was that they began to eat a range of vegetables and fruits that they had previously refused, and were continuing to eat them 6 months later. Another study (Jeffrey et al. 1994) has shown that increased exposure to fruits and salads, together with price reductions, can also increase consumption by adults. An experiment was set up in which the range of fruits for sale in a works canteen was doubled, and 3 additions were made to the salad ingredients offered; consumption rose threefold during the intervention period, but no conclusions regarding the effect of greater exposure to the foods in question alone can safely be drawn as their prices were also reduced by 50%.

The apparent importance of the repeated availability of certain foods in their becoming established in a child's repertoire would suggest that children are likely to develop similar tastes to their parents – assuming, of course, that what parents offer their children is the same as they eat themselves. However, the evidence for this, though significant, is slim (Borah-Giddens and Falciglia 1993). One explanation could be that human tastes develop and change with age. It is one way of understanding the finding of the study by Gibson et al., above, (1998) of mothers and children that, while mothers who ate a lot of fruit had children who did likewise, there was no relationship between mothers' and children's vegetable or confectionery consumption. The children overall ate less vegetables than fruit – indeed it is a common experience to find children refusing vegetables, while many enjoy fruit. A survey I carried out amongst 78 high school pupils (in addition to the study detailed in "A new study") showed that the great majority liked fruit and ate it as a snack and as a dessert, while vegetables were liked by notably fewer. However, over half did enjoy vegetables and three-quarters liked salads. This perhaps illustrates the process of tastes changing with age, for these pupils were older than the children in the study by Gibson and colleagues who ate far more fruit than vegetables. That tastes are not fixed was borne out by about 60 % of the pupils in my school survey, although change was not all in the same direction. Several had ceased to like fruits and vegetables they had formerly been happy to eat, but more had expanded their repertoire, and those who attempted an explanation were unanimous that is was to do with getting older. It seems likely that as children grow up their liking for very sweet things declines (breast milk is sweet), and some develop a greater appreciation of vegetables.

It is not unreasonable to assume that children pick up many ideas about food preparation and provision from their mothers or other food providers as they grow up, and put them into practice as adults, so that eating patterns which are established in childhood are often continued in adulthood; this does not, of course preclude the adding of tastes and adapting of eating habits. Indeed, it has been found that accustomed food comes to be preferred to unfamiliar food, "hence palatability will adapt to the specific foods or drinks to which the individual is frequently exposed" (Conner 1994). This must surely be a good reason routinely to offer children plentiful fruit and vegetables from an early age. But tastes are also susceptible to change in response to new or altered experience – as, for instance, in the case of the children watching the fruit and vegetable video (Fleming et al. 1995), or experimental subjects fed food with lower salt content than they were used to (e.g. Bertino 1982). Herein lies hope that habits can be changed to include more fruit and vegetables.

5.2.5
Summary of Section One

Level of income and general education undoubtedly play a major role in determining the quantities of fruit and vegetables that people consume, and nutritional knowledge is also an important factor. But these influences are moderated by others, such as dominant culinary traditions, commercial practices, the value placed on personal health, time available and for and skills in cooking, social circumstances of eating and, perhaps most of all, personal preferences.

5.3 A New Study

Despite the very real obstacles and the powerful logic of those who argue that the principal barrier to eating plentiful fruit and vegetables is low income, my own experience of people led me to treat this approach with a degree of scepticism, believing that the role of personal attitudes, tastes and values should also be considered. I therefore decided to test this scepticism by a small-scale study. This built on my contacts as treasurer in 1999 and 2000 for a week's summer camp, at which a significant proportion of campers were living on low incomes and who, on the basis of self-assessment, could claim bursaries to help cover the cost of the all-found holiday (full rate £ 60 for adults, less for children according to age). Having this privileged information, I contacted all those who had requested bursaries in the previous 2 years, inviting them to complete a questionnaire about low income and food. I did not explain that my specific interest was fruit and vegetables. Ten agreed to participate, and two offered to pass copies of the questionnaire to others they knew who were living on low incomes. Thus my sample included not only people I knew but also some unknown to me.

Previous experience of researching another everyday activity, that of watching television and videos, and its influence on children's imagination when story-making (Belton 2000), had confirmed my initial position that watching television and the use made of it, must be studied from a holistic perspective, embedding and contextualising it in general everyday experience, rather than isolating it for scrutiny. My investigation of this subject had revealed that children's literary interaction with the screen was indeed part of an intricate matrix of experience, tastes, personal predispositions, and family relationships and values. The approach I took to the present study was therefore very similar, based on the notion that individuals' behaviour in one aspect of life will bear some relationship to their behaviour in others. While human beings are not consistent in what they do, the lives of individuals and families express a certain degree of coherence which characterises them. It was this sort of embedded picture of fruit and vegetable consumption that I wished to elicit, and low income was only one aspect of it.

My sample totalled fourteen households, comprising 20 adults and 29 children: 6 couples with children, 5 single parent families and 3 adults living alone.

Although there was one principal adult respondent for each questionnaire (all but one of whom were women), some questions also invited answers from or on behalf of partners. It was therefore possible to build up a picture both of household patterns and of individual experiences and tastes.

That these were low-income households was in no doubt. 6 included no earner, and all but 3 were in receipt of state benefits such as Income Support, Working Families Tax Credit, Housing Benefit, disability benefit or state pension. One exception was a student living on a loan of less than £ 75 per week and another household had to pay out a substantial sum each week in maintenance to a first family. The group also represented a considerable social mix. Past and present occupations included local government officer, community volunteer, home-maker, literacy and numeracy tutor, barmaid, land worker, nature warden, Open University tutor, overseas development worker, youth worker, civil engineer, library assistant, piano restorer, waitress, builder, lecturer, secretary, teacher, and gardening instructor for adults with learning difficulties.

So what did the survey reveal about the fruit- and vegetable-eating habits of this range of low income households? Overall, their consumption was relatively high. All households but one (a single man) served at least 2 kinds of vegetable on an average a day, 2 households as many as 5. Similarly with fruit, the norm was to have 2 to 3 varieties available; 2 households offered only 1 but 3 offered 4 or more, and 9 claimed to serve fresh, stewed or tinned fruit as a dessert at least once a week. These figures do not, of course, tell us how many servings were actually eaten each day by each member of a household but they do give an indication of the value or emphasis placed on these foods by the main provider of food.

These data suggest that fruit and vegetables were treated by most of these people as an important part of their everyday diet. Other information they gave indicated their attitudes towards other food matters, for the survey also sought, for example, to find out what foods were deemed suitable as snacks in these families and what went into school lunch boxes, as both snacks and packed lunches are frequently occasions for the fatty or sugary items for which healthier alternatives may be easily substituted. Amongst the snacks named, fruit was easily the most popular, with 11 mentions, then biscuits, crisps and dried fruit which were each recorded 5 times. The list offered by participants also included bread, toast, nuts and seeds and home-made cakes and flapjacks. All packed lunches included sandwiches, after which fruit was mentioned more often than any other foodstuff; yoghurt, muesli bars and crisps were mentioned by just under half the respondents. Of the adults, 10 out of 14 claimed that between 5 and 7 of the main meals prepared in their household during a normal week were created from raw ingredients; conversely, ready-prepared meals were not popular. One respondent commented, "It is too expensive to buy convenience foods. Home cooking is usually better. I might use a bought tomato sauce to help with a meal, or vegetarian sausages, or tinned beans."

Evidence that most of this group of people – at least the adults – positively relished fruit and vegetables and healthy food generally came from their answers when invited to name the foods they particularly enjoyed. The most frequently mentioned food was salad, with 5 votes, followed by pasta dishes, fish, vegetable/vegetarian dishes and good/home-made bread with 4 each, then vegetables and cheese with 3. Steak, roast meat, cakes and puddings were mentioned twice each, and chocolate once. It is, perhaps relevant, too, that 17 out of the 20 adults surveyed actively enjoyed cooking, 8 of them sometimes making their own bread. Growing their own food, however, was somewhat less popular. Of the 12 households with a garden, 7 grew some fruit and vegetables, but generally not a significant proportion of what they consumed. The one exception to this was a single, disabled mother of two teenagers who said she was "serious about fruit and veg growing to supplement my income", such that in the summer most of her family's needs were home grown, encompassing about 40 different varieties.

As mentioned above, my research approach would see an individual's or family's attitude to eating fruit and vegetables in practice as one aspect of a larger worldview or set of values, and not purely a function of income or level of education. I therefore included in the questionnaire several questions that aimed to draw out some indication of the attitudes and lifestyles of this particular sample of people who were used to making do on slender means. This would help to construct a wider context in which to see their eating habits. These included questions on how they spent their spare time and any disposable income left after paying for necessities. Responses suggested that most pursued quite active lives, other than in the course of paid work – if employed – following such interests as gardening, bee-keeping, reading, walking, morris dancing, sewing, decorating, keep fit, transporting children, listening to a variety of music, voluntary work, historical re-enactments, nature conservation and crafts. In fact, asking people to show how often their household spent money on a range of possibilities produced a fairly well-defined set of priorities. Respondents indicated regular expenditure as follows: out of the 14 households

- 5 bought newspapers (broadsheets or local papers) daily or weekly
- 6 paid for swimming or some other sporting activity every week
- 4 bought books at least monthly
- 5 paid weekly for children's music lessons or other classes
- 4 bought tobacco (2 daily and 2 weekly)
- 6 bought chocolate or sweets (3 daily, 3 weekly)
- 3 hired videos (2 weekly, 1 monthly)
- Most bought beer, wine or spirits less than once a month

- 3 spent money on CDs, theatre, cinema or a concert monthly
- 6 made a donation to charity every month

These responses suggested a collection of people who were, on the whole, actively involved in life, and made their resources stretch to include cultural and sporting activities. However, this was not necessarily easy. One (single) person, with wide interests and a busy life, but living on Income Support and Housing Benefit, commented, "The food I choose to buy is often more expensive than the cheapest option available, but in terms of nutrition is far more nourishing. The thing about living on a low income that stresses me most is when I run out of money for chocolate." Friends were a source of enjoyment, and one single mother on Income Support spoke of the pleasure of eating good food in good company but the difficulty of being sociable on a tight budget. Another single parent, living on disability benefits, wrote of the distress caused to her and her children by the necessity of telling them at times that they could not invite friends round for meals as there was simply not enough food to offer.

Just as the data yielded by the questionnaires can be interpreted to suggest that people's tastes and habits in eating bear a relationship to their wider lifestyle, attitudes and values, the responses to another question indicated that the basis on which a significant number of the participants in this study chose food to buy took into account more factors than simply cost, taste and health. Clearly price and family likes were the overriding considerations, and healthiness seemed to weigh just as heavily. But other issues came into play too: country of origin, fair trade/unfair trading practices, ethical issues, packaging and other environmental concerns all influenced the purchasing decisions of between 4 and 8 of the 14 principal participants struggling to make ends meet. 5 said they bought organic produce once a week or more; only 4 never did. All this evidence put together indicates that what people value, whether in terms of their own or their family's health or in relation to more global concerns, contributes to the determining of food choices. It supports the notion that diet is more than simply a function of income and education, and the expectation that, apart from the very poorest in society, such as the homeless and some of the young pregnant women referred to above, who have absolutely no financial room for manoeuvre, consumers express personal priorities as well as income and knowledge when choosing what to eat. How values relate to income, education, family history and social class is, however, is an extremely complex matter.

Some of the additional thoughts offered by participants are illuminating of the determination to adhere to a belief in the importance of a healthy diet that some people exercise, low income notwithstanding, and their ingenuity in putting this into practice in the face of considerable financial constraints. One working single parent, whom I shall call Pippa, who had had to give up paying towards a private pension and whose teenage children could not have pocket money, commented that she made her own sauces and soups out of a minimum of ingredients, and would rather spend money on fresh fruit and vegetables than buy convenience foods; and, although she generally used cheaper lines she did use butter and olive oil. Susan, a mother of younger children said, "I find convenience food expensive, I never buy ready meals, or pies, rarely cook-in sauces. I don't think they're a healthy way to eat... I do always put vegetables on the children's plates, even if they're not eaten – sometimes they do." Rosemary acknowledged that producing healthy food is not necessarily straightforward:

Food is the basis of good health but it takes, time, knowledge and equipment to feed children healthily on little money... I balance expensive/inexpensive meals, e.g., roast dinner against lentil soup or sweetcorn chowder... I use the best quality I can, e.g., I use organic farmhouse cheese but only use a little.

This approach had clearly been successful for, although Rosemary's family were entitled to free school meals the children found their quality to be so low compared with what they were used to at home that they wouldn't eat them: "horrible bread, plastic cheese and **squash**", in Rosemary's words. She concluded, "We all enjoy good food and eating together. I have very little pressure for MacDonald's/Microwave chips." Ellen felt equally strongly about the quality of food and wrestled with similar odds, using somewhat different strategies:

I shop at the Co-op which has a 'boring' range of foods so I'm not tempted to overspend, and I work on buying a fairly fixed shopping list. I try to aim to buy in one treat a week, e.g., a melon, tofu, expensive choc biscuits or a different cereal, croissants for breakfast, a white loaf. We try to have one pulse dish a week so vary these too. I try not to stint on good bread, good tea and good fruit and plenty of vegetables. I belong to two LETS schemes [local exchange trading schemes – in which goods and services are traded and no money is involved] which have proved invaluable in upgrading our lives. Excess food from the garden has been used to gain credits to trade for other requirements.

For single Nina, it was a stressful matter of juggling demands. She found that her weekly income was not enough to maintain a healthy diet as well as pay all her bills on time so she made sure that she ate well but often paid her non-priority bills late.

The types of fruit and vegetables served or available every day in participants' households ranged from 3 to 8. This cannot be directly compared with figures from larger surveys, such as the annual National Food Survey, which have attempted to discover actual amounts consumed by individuals daily. However, this figure, together with written comments, suggests that the emphasis placed on fruit and vegetables in most of the households I surveyed was relatively high, and particularly high for people living on low incomes. As well as their income, the questionnaire asked respondents about their education. All levels were represented, from no qualifications at all or very few to two doctorates. Generally, it has been well established that a clear correlation exists between level of education, level of income, quality of diet and quality of health. The small sample represented in my survey, however, only partially reflected the national picture in which fruit and vegetable consumption increases with the level of income, education and social class. In common with the bigger picture, however, there was a discernible, though not wholly consistent, correlation with level of formal education. The extreme example of a respondent who conformed to the general pattern was a single man with no educational qualifications, who apparently rarely ate any fruit or vegetable other than potatoes, and who said that his food purchases were determined by cost rather than choice. But then, he also conformed to the national picture in which less well educated men have been found to be some of the lowest consumers of fruit and vegetables (Fraser et al. 2000).

In addition to factors of education and income, which could be termed gross measures, my study attempted to elicit more subtle indicators of participants' lives, some sense of their interests and cultural activities; in this it is unusual, and the results, though limited, are interesting. For, whereas the correlation between income and fruit and vegetable consumption did not echo that which has been repeatedly obtained for larger populations, and the pattern in relation to educational attainment was somewhat weaker, there did seem to be a discernible relationship between fruit and vegetable consumption and level of leisure-time activity. In other words, there seemed to be some connection between living an active and engaged life and eating relatively generous quantities of fruit and vegetables. This is a question to which I shall return in the third part of this chapter.

In many ways the sample of people represented in this small study, as their range of qualifications and occupations shows, did not generally follow the low income stereotype which is often characterised as educationally and socially deprived and, implicitly, passive. Low income was a common factor among participants, but the study demonstrates that, although acquiring fruit and vegetables on a low income can be a struggle, limited means do not *a priori* rule out the eating of adequate amounts. This would suggest that low income is rarely the sole constraining factor in the quantities of fruit and vegetables consumed. Indeed, conversely, although people living on larger incomes have been shown generally to purchase greater quantities of fresh fruit and vegetables, the average intake in the UK is still only 244 g per day (about 3 portions) rather than the recommended minimum of 400 g (or 5 helpings) (Cox et al. 1998), thus showing that prosperity is no guarantee that people will eat the recommended quantities of these foods.

While large-scale quantitative surveys show that income and education certainly do exert a powerful influence on diet over a large population, even a small-scale survey of a qualitative nature has the potential to show how personal values and worldviews can interact with and moderate the general effects of income and level of formal qualifications. **Broad-based** qualitative study can probe beneath the general patterns generated by quantitative research and highlight individual differences, adding fine detail to what can be discovered, and thus expand understanding to include some of the more subtle factors which affect human attitudes and behaviour.

An extension to the kind of questions posed by my questionnaire which could be interesting to pursue in order to enhance understanding of personal motivation regarding diet, would be to investigate whether any relationship exists between an individual's perception of life in relation to nature and what they eat. That is, do those with a spiritual or philosophical orientation which entails a perception of themselves as being in a dynamic relationship with the natural world have a distinctly different approach to food and eating from those whose orientation is towards, say, the technological or commercial aspects of life? Is there a difference in this context between those one might characterise at the extremes as the "earthy" and the "worldly"?

Four themes emerged clearly from the responses that were given in this study which seem to be key issues in the whole question of the eating of fruit and vegetables. They are the importance of enjoying these foods, and of enjoying cooking, the significance of valuing a healthy diet, and the suggestion of a connection between healthiness of diet and an active, socially engaged life. This fourth question is one which deserves more attention, and I shall address it further now, in the final part of the chapter.

5.4 The Role of Psycho-social Health in Influencing Fruit and Vegetable Consumption

Healthy eating, and specifically the consumption of fresh fruit and vegetables, is clearly the result of a complex interaction of experience, education, income, taste, values, culture and other influences. However, there are indications that another important but much more subtle factor, may also contribute to individuals' levels of fruit and vegetable consumption. That factor is psycho-social health, or how people **feel** in relation to their social world. Conventional nutritional thinking has put much emphasis on the impact of diet on physical health, and justifiably so. However, reversing this emphasis allows one to consider instead the influence of health on diet, for 'health' is a concept that is now acknowledged, by the lay public as well as medical experts, to encompass more than just the physical (Blaxter 1990).

Findings of Mildred Blaxter's Health and Lifestyles survey which support this shift of emphasis demonstrate a strong association at all ages between diet and social health, and indicate the major role that social support and integration plays in general health and the perception of being healthy; those with many activities were found to be unlikely to call their lives unhealthy. The concept of social health is not tightly defined by Blaxter but denotes a sense of mental and emotional wellbeing, confidence, and enjoyment of life. Consumption of fruit and salad was clearly negatively associated in the survey both with physical illness and with psychosocial malaise, as identified by symptoms such as depression, strain, worry and sleep disturbance. Indeed, a major conclusion drawn from this study by Blaxter was that social circumstances, as well as physical ones, play an important part in determining levels of physical health. These Health and Lifestyles data indeed support the tendency I noted amongst the participants in my own study, of those whose lives seemed more active and socially engaged to eat more fruit and vegetables. In this sense, fruit and vegetables could be regarded as the antithesis of comfort foods which help to bring consolation for social or emotional problems. Further hints of a connection between social engagement and healthy eating are provided by the impression given by comments of participants in 'food projects' that eating more healthily was, in the course of participating in such projects, tied up with an enjoyment of sociability, and by a study in the US (Sandersphillips 1994) which found that low-income black women and Latinas were significantly more likely to have healthy eating habits if they enjoyed higher levels of social support. However, the significance of emotional states has not so far been recognised as an area for thorough investigation in relation to levels of fruit and vegetable consumption.

Richard Wilkinson, a sociologist of health, has assembled a large volume of empirical data which show that health, (rates of both disease and death), is fundamentally affected by psychosocial factors, in other words, that people's **feelings** in response to either chronic or acute social or emotional situations exert a powerful influence on their **physical** health. He explains the connection in terms of many of the biological processes that lead to illness being triggered "by what we think and feel about our material and social circumstances" (Wilkinson 2000 p. 2). He gives numerous examples of evidence that people further up the social hierarchy tend to be healthier than those lower down. This might not seem surprising, but Wilkinson's explanation is not the obvious one that the correlation is a simple function of higher income and better education:

Underlying these health inequalities are some intensely social factors that exert a strong negative influence on health. They include the direct effects of subordinate social status itself (as distinct from the effects of the poorer material circumstances that normally accompany it) of friendship, of social networks, and lack of autonomy and control (ibid p. 3).

As an illustration Wilkinson cites research which shows that junior civil servants, at the bottom of their particular pecking order, have four times the rate of heart disease suffered by senior civil servants (Marmot et al. 1991). More generally, chronic stress, caused not only by poverty, but also by psychosocial insecurity resulting from such experiences as bereavement and threat of redundancy, have been shown to have marked physiological effects such as suppression of the immune system.

What matters to health, Wilkinson says, is not absolute income and living standards but relative income and social status. In environments where there is less economic and social disparity people are more likely to trust each other and are less hostile and violent towards each other. Thus, in the developed world, life expectancy is highest not in the richest societies, such as the US, but in those with the least inequality, such as Japan. Therefore,

The importance of income distribution implies that we must explain the effect of low income on health through its social meaning and implications for social position rather than through the direct physical effects (Wilkinson 1996, p. 176).

5.4.1 The Significance of Social Class

The insight that positive psycho-social health possibly disposes people to eat more fruit and vegetables could help to explain why middle class people have a generally higher consumption of fruit and vegetables than working class people, beyond their usually enhanced buying power and level of education. As well as the advantage of elevated social status, they also have the benefit of the intrinsic satisfaction bestowed by learning (Putnam 2000). Both factors contribute to psycho-social health in their own right, independently of the contribution to material wellbeing they usually entail. The emotional support of family and friends, and involvement in different aspects of civic life, which are other important determinants of psycho-social health, however, are perhaps less closely associated with social class than income and education traditionally have been, varying on a more personal level.

Wilkinson points out that, of the many possible pathways which link emotions and physical health, it is not yet understood which are important. One potential explanation for the detrimental effect of poor psycho-social health on physical health is that social isolation has been shown to have measurable biochemical effects on the body (Putnam 2000). Some evidence presented in this chapter, seems to suggest that another possible connection is the greater likelihood of someone eating plenty of fruit and vegetables if they enjoy a relatively high level of psycho-social health. In other words, psycho-social wellbeing may facilitate the eating of relatively large quantities of fruit and vegetables which, in turn, benefit physical health by nutritional means.

Positive psycho-social health is likely to encompass a degree of personal security, confidence, fulfilment, self-esteem and warm and supportive relationships, and these attributes are likely in themselves to engender some sense, albeit implicit, of a positive future. The individual's self-perception in relation to the future could be significant in the shaping of their eating patterns. It is important to understand this for, as Calnan and Williams' (1991) study cited above showed, while middle class people are significantly more likely deliberately to eat in a way that will guard against a potential future risk to health, working class people tend to adopt healthier eating habits only in response to actual manifestations of ill-health in the present. This class difference suggests that positioning of the self with respect to the future may play a key part in food choices. I will now offer two ways of understanding the relationship.

One route to understanding class-based perceptions of the future is suggested by the investigation of Basil Bernstein into the forms of language used by different social groups. According to Bernstein (1973), our fundamental conceptualising and thinking processes are shaped by the way we learn to use language in the home. His analysis of material gathered during many years of observing and recording spoken language, led him to conclude that the language of middle class and "lower working class" (unskilled and semi-skilled) people constructs markedly different relationships between people and objects. This is the case irrespective of level of measured intelligence. According to him,

> *The language-use of the middle class is rich in personal, individual qualifications, and its form implies sets of advanced logical operations ... The child in the middle class ... grows up in an environment which is finely and extensively controlled; ... The more purposeful and explicit the organisation of the environment with reference to a distant future, that is the greater the rationality of the connections and inter-relations between means and distant ends, the greater the significance of objects in the present* (pp. 49–50).

In contrast, Bernstein suggests, the working class family structure is less formally organised in relation to the development of the child. Its values

> *do not give rise to the carefully ordered universe spatially and temporally of the middle class child ... The specific character of long-term goals tends to be replaced by more general notions of the future, in which chance, a friend or relative plays a greater part than the rigorous working out of connections. Thus present, or near present, activities have greater value than the relation of the present activity to the attainment of a distant goal* (pp. 53–54).

If Bernstein's view is even partially correct, the implications for choice of diet in relation to long-term health are obvious.

A quite different perspective on class divergence emerges from a recognition that the concept of 'health', even when applied to oneself, is open-ended, unbounded by time. There is a sense in which it is not possible to know whether one has achieved the goal of healthiness, as one cannot know what diseases or conditions may affect one in the future. Therefore deliberately living healthily is an act of faith, a process which entails taking or avoiding certain actions on a more or less daily basis in the hope that these will stave off an indeterminate impairment to one's well-being at an unspecified point in the future. Such behaviour surely requires of a person that they feel a tangible stake in the future, and a belief in the long-term efficacy of their immediate actions.

A second way, therefore, of looking at the greater propensity of middle class people to eat more adequate quantities of fruit and vegetables, is to say that those who enjoy the advantages usually taken to characterise being middle class, are likely to take their future for granted, and are thus likely to be motivated to invest in that future with health-related behaviours. For those, on the other hand, whose present existence is more of a financial struggle and less emotionally satisfactory, in the sense of feeling socially and educationally inferior, living in the present moment is likely to predominate over considerations relating to a time far ahead. For them any potentially health-enhancing behaviour, whose uncertain benefits would be experienced only in a distant future, is much less likely.

5.4.2
Social Upheaval and the Need for Comfort and Convenience

Since Bernstein was working, class divisions have become less well defined – hence 8 of my 20 low income participants regarded themselves as middle class, 5 as classless, and only 7 as working class. Income level is now less closely associated with social class than it was thirty years ago, and nowadays low income is the result of a variety of circumstances, not simply a lack of educational qualifications and thus of employability in well paid jobs. It can arise, for instance, out of a spell of unemployment, or, increasingly commonly for women, divorce; indeed, divorce was a factor in half the households who participated in my survey. In other words, living on a low income is now experienced by many people who are not working class in Bernstein's terms – and whose lives are not impoverished in a broader sense. The 'modern' era in which Bernstein worked, with its relatively clear-cut objectives and direction, has been superseded by the uncertainties of the post-modern world. This new world is characterised by the insecurities of new developments such as the end of manufacturing industry, short-term contracts, the common experience of family break-up and the fluidity of what used to be dependably fixed: the divide between, for instance, day and night, male and female, work and leisure.

Thus changes in the social and economic world are affecting not only the working class. The security of the middle class has also been eroded; indeed the whole fabric of society has begun to lose its once distinct form. Robert Putnam (2000) has recently attested to the late twentieth century collapse of community in America, where civic disengagement and the dissolving of social connections has coincided with an increase in rates of depression and suicide. In a parallel study Oliver James (1998) has described how the disintegration of social structures and the economic pressures experienced in Britain in the second half of the twentieth century have produced an "emotionally toxic" society. If there is a real association between psycho-social health and eating habits, the powerful and damaging social trends described by Putnam and James may go some way to explaining why even most of those who earn enough and are well educated enough to make sure they consume the recommended daily quantities fruit and vegetables fail to do so. This, combined with long working hours which compete with time for cooking, the powerful processed food industry and the prevalent prizing of "convenience" and speed are all likely to bestow more value on the "product" to be eaten than on the personal "process" of creating it.

It is often claimed that the reason that people living in poverty do not eat enough fruit and vegetables is that "in terms of calories per penny, chocolate is a better deal than carrots" (Lobstein 1997). It is undoubtedly true that one can buy far more calories for a modest sum at the baker's or sweet shop than one can at the greengrocer's. However, perhaps this is not the only reason for choosing foods high in fat, sugar and carbohydrates, for literature consistently demonstrates that individuals experiencing emotional distress, especially depression, report a craving and preference for sweet carbohydrate/fat rich foods (Christensen 2001). This seems to be the result of an enhancement of mood following consumption of such foods. Wurtman and Wurtman (1995) explain the physiological mechanism involved thus: carbohydrate consumption, acting via insulin secretion and the 'plasma tryptophan ratio', increases serotonin release, and serotonin release is involved in mood control. According to them, it has been observed that many patients suffering depression, seasonal affective disorder and premenstrual syndrome learn to overeat carbohydrates, particularly snack foods like crisps and pastries, because it makes them feel better, and quickly so. In this way they are using certain foods as though they were drugs. Likewise, people have been found to respond to periods of stress induced by unusually high workload or unemployment, by increasing their consumption of foods high in fat and cholesterol (McCann et al. 1990).

A note of caution must be sounded, for another study, by Young (1991), suggests that lowered mood caused by low levels of brain serotonin is probably not involved in any simple way in carbohydrate craving, and that carbohydrate is possibly able to affect mood by mechanisms other than through the alteration of brain serotonin. But it is undeniable that the well known attraction of sweet, filling and fatty foods to people when they are feeling down is has resulted in the coining of the expression "comfort foods". James (1998), a clinical psychologist, suggests that there is an epidemic of low levels of serotonin in the UK population, induced by feelings of unhappiness and personal inadequacy in a fiercely competitive society. The huge increase in the consumption of snack foods, referred to earlier, could therefore represent self-medication with food on a large scale. This social development affects the affluent and successful as well as the poor. High consumption of comfort-inducing foods may reduce levels of fruit and vegetable consumption in 4 ways: by supplanting them with other types of food (for instance a chocolate bar in stead of an apple as a snack), by filling the stomach so that vegetables or fruit seem an unnecessary and undesirable part of a meal, by wholly displacing a proper meal, or by using up the money available for food as a whole. Depression and a diminished social role can also cause a person simply to make less effort to prepare meals, or to eat less altogether; as one of my participants commented, "I ate and prepared food which was more organic, grew more vegetables and owned an allotment when my family was larger. My daughter and I eat more convenience foods now. I no longer bake my own bread or cakes or make my own soup. I eat less now due to depression and when I am alone." It is possible that what appears to be low morale on a societal scale, together with the prevalent desire for convenience and speed in assembling food, could orientate large numbers of people towards carbohydrate- and fat-rich, processed, foods and away from fresh and often messy or time-consuming fruit and vegetables.

5.5 Summary and Conclusions

There are a number of conditions which must be satisfied if individuals are, in the main, to eat appropriate quantities of fruit and vegetables to optimise long-term health. Firstly, individuals must be in possession of an adequate income, and there is no doubt that urgent steps need to be taken to eradicate poverty in Britain. Secondly, good access to a range of fresh, good quality produce is essential if people are to be physically able to meet their needs in terms of fruit and vegetables. Thirdly, a culinary culture in which vegetables and fruit play as prominent a role as other basic foods such as bread and cheese, as they do in Mediterranean countries, for instance, encourages consumption. A shift in this direction could be proactively encouraged by measures such as firm government minimum standards with regard to catering in state-run organisations such as schools, hospitals and residential homes. In addition, children could be offered free fruit at school in the same way that they were once provided with free school milk. Fourthly, the ability to cook simple meals from fresh ingredients should be regarded as an essential life skill, and taught to every pupil at school. Fifthly, although, there will always be some who lack the cognitive skills to process complex diet and health messages (Lambert 2001) and those whose ears will be closed to such information and advice, efforts must continue to inform the public of the important benefits of eating sufficient quantities of fruit and vegetables and what those quantities are. The majority of the population of Europe and the US, despite strenuous efforts on the part of health educators, are still far from clear about the five-a-day recommendation (e.g. Lambert 2001). Sixthly, the greatest motivator for eating fruit and vegetables is the positive enjoyment of doing so. Simply being exposed to them on a daily basis, attractively presented and well cooked, and seen to be eaten by appropriate role models, seems the means of cultivating a taste for them most likely to achieve success, particularly for children. All the initiatives I have suggested here would be likely to assist in making fruit and vegetables a more attractive and prominent part of the popular diet.

Lastly, and most controversially, is the question of the role of psycho-social health. More research is required into the influence of social and emotional health on food choices and eating habits. A range of evidence presented here points to the likelihood of a relationship, but further investigations are needed. The diverse research assembled and collectively interpreted by Wilkinson (1996, 2001) and by Putnam (2000) strongly suggests that personal social integration and social status are critical factors in determining an individual's level of psycho-social health. These are heavily influenced by levels of income and education and the degree of choice thus bestowed. But psycho-social health is also crucially affected by the quality of human relationships that individuals experience. The quality of human interactions, whether of a personal, public or institutional nature, could often be greatly enhanced by a better understanding of the fundamental contribution of emotions to human behaviour and responses and of how emotions are generated. A more emotionally literate society would be able to design and implement policies and practices at all levels which would help to create a much healthier society in the broadest sense.

As this more holistic approach to the organisation of society has yet to be widely grasped an illustration may clarify its nature and its relevance here. One means by which many people living on a low income, especially those without paid work, have found to help themselves both materially and socially is to join a LETS scheme, a local exchange trading system in which individuals trade time and skills for tokens and no money changes hands. Belonging to such a scheme has been found to enable vulnerable people to maintain labour skills, self-esteem, social interaction and independence. These arrangements are looked on positively in Australia, New Zealand and Ireland, but in the UK participants on benefits are lucky if their activities and "earnings" are not penalised through "benign neglect" (Croall 1997). One example of the kind of way in which UK government policy could take into account the emotional impact of official policies would be to abandon the arbitrary, inconsistent approach taken by Department of Social Security officers to the non-monetary earning potential for participation in LETS which results in participants putting their benefits in jeopardy. If a more empathetic and constructive, less punitive attitude were adopted by officialdom to those living on a low income, the stigma of poverty and the associated inferior status accorded to those living on a low income would be ameliorated, social cohesiveness enhanced and division diminished.

This final thought has departed a long way from the narrowly focused approaches hitherto taken to the issues involved in fruit and vegetable consumption, and suggests that it is now time to develop a much broader and socially contextualised understanding of eating behaviour. The quantities of fruit and vegetables that people routinely eat are determined by an interaction of far more complex and subtle factors than simply the income, general level of education or health information they receive. It seems likely that a range of social factors which affect wellbeing also influence diet. Social development that is founded on a radically improved understanding of the fundamental role and origins of human emotions seems to be the most promising way of building a healthier society in every sense, one whose members will perhaps be less inclined to eat quick, convenient, comfort-inducing food and more likely to enjoy abundant fresh fruit and vegetables. But such a society will take time to build. In the meantime, it would be possible to design and implement practical policies which would quickly make the eating of healthy quantities of fruit and vegetables both easier and more inviting for a large proportion of the population.

Acknowledgements

The author would like to thank Dr Fiona Poland and Dr Katharine Kite for their valuable comments on a draft of this chapter, Dr Julia Lawton for permission to use her unpublished work, and the participants in the study reported.

References

Ajken I, Fischbein M (1980) Understanding attitudes and predicting social behaviour. Prentice Hall, New Jersey

Axelson ML, Federline TL, Brinberg D (1985) A meta-analysis of food- and nutrition-related research. J Nutrition Ed 17: 51–54

Belton T (2000) The 'Face at the Window' study: a fresh approach to media influence and to investigating the influence of television and videos on children's imagination. Media, Culture and Society 22: 629–643

Bernstein B (1973) Class, codes and control. Vol 1: Theoretical studies towards a sociology of language. Paladin, St Albans

Bertino M, Beauchamp GK, Engleman K (1982) Long-term reduction in dietary sodium alters the taste of salt. Am J Clinical Nutrition 36: 1134–1144

Blaxter M (1990) Health and lifestyles. Routledge, London New York

Borah-Giddens J, Falciglia GA (1993) A meta-analysis of the relationship in food preferences between parents and children. J Nutr Ed 25: 102–107

British Heart Foundation (1999) Coronary heart disease statistics. British Heart Foundation, Oxford

Brug J et al. (1995a) Psychosocial determinants of fruit and vegetable consumption. Appetite 25: 285–296

Brug J, Debie S, van Assema P, Weijts W (1995b) Psychosocial determinants of fruit and vegetable consumption among adults: results of focus group interviews. Food Quality and Preference 6: 99–107

Calnan M, Williams S (1991) Style of life and the salience of health: an explanatory study of health related practices in households from differing socio-economic circumstances. Sociology of Health and Illness 13: 506–529

Carlowe J (2001) Observer Magazine 11 March: 25–26

Christensen L (2001) The effect of food intake on mood. Clinical Nutrition 20: 161–166

Committee on Medical Aspects of Food and Nutrition Policy (1998) Nutritional aspects of the development of cancer; Report of the Working Group on Diet and Cancer of COMA

Conner MT (1994) An individualised psychological approach to measuring influences on consumer preferences. In: MacFie HJH, Thompson DMH (eds) Measurement of food preferences, pp 167–201. Blackie Academic and Professional, London

Co-operative Wholesale Society (2000) Blackmail. CWS, Manchester

Cox DN, Anderson AS, Lean MEJ, Mela D (1998) UK consumer attitudes, beliefs and barriers to increasing fruit and vegetable consumption. Public Health Nutrition 1: 61–68

Croall J (1997) LETS act locally: the growth of local exchange trading systems. Calouste Gulbenkian Foundation, London

Dallison J, Lobstein T (1995) Poor expectations: poverty and undernourishment in pregnancy. The Maternity Alliance and NCH Action for Children, London

Department of Health (1998) Our healthier nation; a consultation paper. The Stationery Office, London

Earland J, Ibrahim SO (1995) Maternal employment: does it influence feeding practices during infancy? Abstract Fourth Food Choice Conference University of Birmingham UK 1995. Appetite 24: 268

Fleming PFJ, Horne AJ, Dowey AJ, Lowe CF (1995) Changing food preferences in children. Abstract Fourth Food Choice Conference University of Birmingham UK 1995. Appetite 24: 266

Food Commission (2000) Food Magazine October 2000 (http://ourworld.compuserve.com/homepages/foodcomm/oof.htm)

Fraser GE, Welch A, Luben R, Bingham SA, Day NE (2000) The effect of age, sex, and education on food consumption of a middle-aged cohort – EPIC in East Anglia. Preventive Medicine 30 (1) 26–34

Gibson EL, Wardle J, Watts CJ (1998) Fruit and vegetable consumption, nutritional knowledge and beliefs in mothers and children. Appetite 31: 205–228

Gordon D, Adelman L, Ashworth K, Bradshaw J, Levitas R, Middleton S, Pantazis C, Patsios D, Payne S, Townsend P, Williams J (2000) Poverty and social exclusion in Britain. Joseph Rowntree Foundation, York

Gowdridge C, Williams AS, Wynn M (eds) Mother Courage: letters from mothers in poverty at the end of the century. Penguin, London

Hupkens CLH, Knibbe RA, Drop MA (1995) Class differences in food rules in nuclear families: a cross-cultural survey. Abstract Fourth Food Choice Conference, University of Birmingham UK 1995. Appetite 24: 269

James O (1998) Britain on the couch: treating a low seratonin society. Arrow, London

Jeffrey RW, French SA, Raether C, Baxter JE (1994) An environmental intervention to increase fruit and salad purchases in cafeterias. Preventive Medicine 23: 788–792

Kuhl J, Beckmann J (1985) Action control: from cognition to behaviour. Springer Verlag, Heidelberg

Lairon D (1999) Mediterranean diet, fats and cardiovascular risks: what news? Invited Commentary. Brit J Nutrition 82: 5–6

Lambert N (2001) Food choice, phytochemicals and cancer prevention. In: Frewer LJ, Risvik E, Schifferstein H (eds) Food, people and society: a European perspective of consumers' food choices. Springer, Heidelberg, pp 131–154

Lawton J (in preparation 2001) Colonising the future: temporal perceptions and risk-related health behaviours across the adult lifecourse, Research Unit in Health, Behaviour and Change. The University of Edinburgh

Lobstein T (1997) Myths about food and low income. National Food Alliance, London

Marmot MG, Davey Smith G, Stansfield S, Patel C, North F, Head J (1991) Health inequalities among British civil servants: the Whitehall II study. Lancet 337: 1387–1393

McCann BS, Warnick GR, Knopp RH (1990) Changes in plasma lipids and dietary intake accompanying shifts in workload and stress. Psychosomatic Medicine 52: 97–108

McGlone P, Dobson B, Dowler E, Nelson M (1999) Food projects and how they work. Joseph Rowntree Foundation, York

National Association of Farmers Markets. www.farmersmarkets.net/

Ministry of Agriculture, Fisheries and Food (1999) National food survey 1998. The Stationery Office, London

Putnam RD (2000) Bowling alone: the collapse and revival of American community. Simon and Schuster, New York

Rappoport LH, Peters GR, Huff-Corzine L, Downey RG (1992) Reasons for eating: an exploratory cognitive analysis. Ecology of Food and Nutrition 28: 171–189

Roos G, Johansson L, Kasmel A, Klumbiene J, Prattala R (2000) Disparities in vegetable and fruit consumption: European cases from north to south. Public Health Nutrition 4: 35–43

Sandersphillips K (1994) Correlates of healthy eating habits in low-income black women and Latinas. Preventive Medicine 23: 781–787

Severin WT, Tankard JW (1992) Communication theories: origins, methods and uses in the mass media. 3rd edn. Longman, London

Sellen D, Tedstone A, Frize J (2000) Young refugee children's diets and family coping strategies in East London. Final Report. London School of Hygiene and Tropical Medicine, London

Shepherd R, Sparks P (1994) Modelling food choice. In: MacFie H and Thompson D (eds) Measurement of food preferences. Blackie Academic and Professional, London, 167–201

Steptoe A, Pollard T, Wardle J (1995) Development of a measure of the motives underlying the selection of food: the food choice questionnaire. Appetite 25: 267–284

Walker B (1997) Understanding boys' sexual health education and its implications for attitude change. Centre for Applied Research in Education, University of East Anglia, Norwich

Webster J (1998) Food poverty: what are the policy options? National Food Alliance, London

Wilkinson R (1996) Unhealthy societies: the afflictions of inequality. Routledge, London

Wilkinson R (2000) Mind the gap: hierarchies, health and human evolution. Weidenfeld and Nicolson, London

World Cancer Research Fund and American Institute for Cancer Research (1997) Food, nutrition and the prevention of cancer: a global perspective, WCRF and AICR, Washington DC

Wurtman RJ, Wurtman JJ (1995) Brain seratonin, carbohydrate craving, obesity and depression. Obesity Research 3: 477–480

Young SN (1991) Some effects of the dietary components on brain-seratonin synthesis, mood, and behaviour. Canad J Physiology and Pharmacology 69: 893–903

Zizza C et al (2001) Significant increase in young adults' snacking between 1977–78 and 1994-96 represents cause for concern. Preventive Medicine 32 (4): 303–310

Novel Foods: The Changing Regulatory Response

6

DEREK BURKE

6.1 Introduction

Policy and practises regarding the regulation of novel foods have been dominated in the last 10 years by the newfound ability to modify the genetic material of plants through recombinant DNA technology. This has found many interesting and important applications for it offers food and feed products with novel traits. The first generation of genetically modified (GM) crop plants were generally developed for such traits as increased resistance against pests or virus diseases and enhanced tolerance to herbicides. More recently, it has become possible to modify metabolic pathways of plants to improve such traits as nutritional characteristics, or safety, for example by lowering the content of allergens in crops.

All such new food products must be tested for safety for the consumer. Advanced nations have all had, for a number years, appropriate systems for the testing of food additives and contaminants, but it became clear in the middle 1980s that the approaches that had been used up till then were not readily applicable to the safety evaluation of whole foods or of major food components. In particular, the limitations of conventional toxicological studies became apparent when animal feeding studies were used to assess the safety of irradiated foods.

Safety issues arising from the presence of additives and from any contamination with micro-organisms or from other causes had been dealt with by the use of appropriate expert advisory committees. The detailed procedures that have been used in North America and Europe vary from country to country but what they have in common is the harnessing of the best technical and scientific advice to give clear guidance to government ministers, who will then have to make the appropriate decision as to whether the product in question is safe enough to be marketed.

6.2
The Situation in the United Kingdom

In the United Kingdom, the responsibility for advising ministers over all issues connected with novel foods and processes, including products from genetically modified crop plants, lies with the Advisory Committee on Novel Foods and Processes (ACNFP), a committee set up in 1988, as a successor to a committee which was concerned with the safety of irradiated foods. As such, the ACNFP is one of the network of independent advisory committees that advise the British government.
The Committee is an independent body of experts whose remit is:

to advise the central authorities responsible, in England, Scotland and Northern Ireland respectively on any matters relating to novel foods and novel food processes including food irradiation, having regard where appropriate to the views of relevant expert bodies.

Although appointed by Government, the committee is *not* made up of civil servants, although of course they play an important role in preparing the documents, and in carrying out the decisions. The Committee is currently made up of 15 experts in a number of different areas – genetic manipulation, microbiology, toxicology, nutrition, plant breeding, and including a consumer representative and ethical advisor. They are drawn mainly from universities and research institutes, with valuable expertise drawn from some companies. Members are appointed for their own expertise, and *not* as representatives of their institution or profession. Although it is an *advisory* committee, its advice has always been taken very seriously by Ministers and in my experience, they have invariably accepted it. There was very little pressure from interested parties during my period as chairman – pressure that was always counter-productive.
I was chairman of the Committee from its formation in 1987 for 9 years. During much of that period I was Vice-Chancellor of the University of East Anglia, although previously I had had a career in the biological sciences, working on the molecular biology of virus multiplication and on the antiviral protein interferon for a number of years, and I also had some practical knowledge of gene cloning. Rules about conflict of interest have always been strict; and any interest *had* to be declared before discussion of the item. This was recorded in the minutes, and the member left the room. Once a year, in the Annual Report, all members' interests are recorded, while the chairman was not permitted to have any such interests at all in any area of the food industry. We were not paid for our work, although we did receive travelling expenses and a modest attendance fee.
Throughout my period as chairman we met 4 times a year, usually for 3 or 4 hours, although over this period, an increasing amount of our work was

done by correspondence between meetings, particularly as the new EU novel food regulations came into effect. Initially I spent about half a day a week on ACNFP business, but this grew as the storm over GM foods developed to about 2 days per week. Luckily I had retired by this time!

Over the first 9 years of the Committee's work, we approved just over 50 products, 16 of which were products of genetic modification. A relatively small number of applications were refused outright but a number of others were, after we had asked for more specific information, not pursued by the applicant. It might be useful to briefly summarise those that were approved, as follows:

Foods obtained through genetic modification

- enzymes such as chymosin, from three different sources
- oils from genetically modified crops such as rape, maize and soya
- tomatoes and tomato paste
- herbicide resistant crops such as soya and rape

Non-genetically modified novel foods

- mycoprotein
- speciality oils
- modified fats and sugars
- novel cereals such as lupin
- microorganisms such as Lactobacillus

However, it has been our experience that most of the public attention has concentrated on the foods obtained through genetic modification.

We worked initially on a case-by-case basis, brainstorming in the committee to try and identify anything that could possibly go wrong. We later systematised this process into a series of decision trees, consulting widely as we developed this process. We have also used the Hazop analysis process, developed in the chemical industry as a systematic way of assessing risk, to check the rigour of our decision trees. It gave very similar results.

We found the concept of *substantial equivalence* especially useful. This is a concept developed initially by OECD and subsequently endorsed by WHO. It involves a comparison with the non-modified food by all the criteria that are available. In this way we could look not only for any direct changes, but also for any *indirect* changes, caused for example by any perturbing effect on metabolism. According to this approach, evaluation is based on a comparison of the GM crop plant or its product with the conventionally bred counterpart, assuming that these foods and feeds have a long history of safe use. This concept has been criticised (Millstone et al. 1999). But the recently held FAO/WHO Expert Consultation has underlined the value of the substantial equivalence approach and has emphasised that the principle is only a guiding tool to identify similarities and differences between a genetically modified crop and its traditional counterpart (FAO 2000). Differences are subsequently assessed as to the extent to which they might affect the safety of the product. The consultation concluded that a comparative approach focusing on the determination of similarities and differences between a genetically modified food and its conventional counterparts aids in the identification of potential safety and nutritional issues and is the most appropriate strategy for the safety and nutritional assessment of genetically modified foods. The consultation also concluded that there were at present no alternative strategies that will provide a better assurance of safety for genetically modified foods. But the usefulness of this principle is not as a safety assessment in itself; it is a prelude to that assessment.

The systematic application of the concept of substantial equivalent to genetically modified foods, which has been developed using currently available technology, is intended to lead to the identification and subsequent safety evaluation of any differences from the comparison material. Future development of genetic modification technology will allow much more complex genetic modifications, involving multiple gene transfers between species, with the potential to influence more metabolic pathways within the organism. This will almost certainly increase the potential for unexpected effects of a modification and raises the possibility that comparison of specific, targeted, components may not be adequate to detect all unexpected changes.

In these instances, non-targeted profiling techniques will almost certainly have to be used to detect differences between genetically modified foods and the counterpart. Such techniques, which include genomics, proteomics and metabolomics, have the potential to compare the level of many components. However, they are still in their infancy and they have yet to be validated for this purpose; furthermore the statistical analysis of a vast amount of data produced by such techniques will be complex. Even having identified differences between the GM food and its counterpart, in the form of a particular profile or pattern of peaks, these will need to be attributed to specific components and the food safety consequences assessed. This will not be an easy task.

In the ACNFP we invariably needed to go back to the company for further information, including raw data if appropriate, and further experimental work involving, for example field trials of new plant cultivars or toxicity testing, was often needed. In this way we conducted a full scientific risk evaluation. Once we were satisfied, we then *recommended* to Ministers, who after they had accepted our advice, then issued Government approval. Similar procedures have been used to establish guidelines, for example, on the use of antibiotic resistance markers, and the UK practise has been of considerable assistance in framing the new Regulations in Brussels.

This situation changed when the EC Novel Food regulations came into effect on 15 May 1997. The European Community established a legal framework for the market introduction of products derived from GM plants in Regulation 258/97/EC – on novel foods and novel food ingredients (EU 1997a). The accompanying guidelines for safety evaluation are based on the principle of *'substantial equivalence'* described above (OECD 1993, EU 1997b). The Regulation defines a novel food as 'a food that has not hitherto been used for human consumption to a significant degree within the Community'. The Commission has accepted that foods that have already been sold extensively in one or more Member States fall outside the scope of the Regulation.

From that date companies wishing to market a novel food in the EU were required to submit an application to the competent authority in the Member State where they first intend to market their product. Following acceptance of the application, the competent authority is required to complete an initial safety assessment and forward it to the Commission within 90 days. The Commission then copies it to the other Member States for their comments, which have to be made within 60 days. If the initial assessment is favourable, and no objections are raised by other Member States, then the product can be marketed. If objections were raised, these are then referred to the EC Standing Committee for Foodstuffs for final agreement, and if necessary, to the Scientific Committee for Food.

6.3 The Changing Climate for Regulation

But how did our thinking change over the first 9 years of the Committee? We used to think, we scientists, that all we had to do was to decide whether a novel food or process was safe or not, and that the consumer would accept what we, the experts, had decided. We should have known better; the refusal of the consumer to accept food irradiation, a process I believe to be perfectly safe, should have made us think again. But we had to learn the hard way! We had not grasped at that time the importance of consumer perception or understood the very different way the consumer sees risk. Nor had we realised what I believe lies behind these changes; the growth of a significant public ambivalence to this new technology, and the risks involved.

We should of course treat all new technologies with caution, for that is what regulation is about, but it does seem that the gap between technological capability and public acceptability is growing.

6.4
A Case Study – An Early Set Back, and the Committee's Response

In late 1988, we were asked to approve the use of a genetically modified baker's yeast. The yeast, which had been developed by Gist-Brocades, involved changing the ability of the yeast to utilise maltose by introducing two genes from a similar, but not identical yeast. Increased utilisation of maltose led to more rapid production of carbon dioxide, and to an increase in the rate at which the bread rose, so offering a commercial advantage. This seemed to us at the time a good case with which to start. After all, the naturally occurring yeast mating process could have brought about such a genetic change.

We could not see any problem, and said so, and in early 1990 a brief press release appeared which announced, "The product may be used safely".

The press reactions varied widely. For example:

"Genetic yeast passed for use" (The Times),
"Man-made yeast raises temperature" (The Independent),
"Bionic bread sales wrapped in secrecy" (Today),
"Mutant yeast is half baked way to slice up nature" (Today),
"Are the boffins taking the rise out of bread?" (The Star).

While the Consumers' Association said "We think all genetically altered foods should be labelled".

So much public concern was aroused that the product has never been used; a harbinger of the problems that were to be encountered later.

We reacted to this by – at the suggestion of MAFF (the then Ministry of Agriculture, Fisheries and Food) – gathering together, for a weekend conference in October 1990, representatives of all the groups who might be able to help us avoid this problem in future. The group included scientists, industrialists, social scientists, a philosopher, a theologian, consumers and members from Green Alliance and Genetics Forum. As a result we made 8 recommendations to Ministers:

- broadening the membership of ACNFP to include a consumer representative and an ethical advisor,
- allowing observers into ACNFP meetings (the only one not proceeded with, although more recently advisory committees have started to meet in public),
- improving access to information (press releases before meetings to briefly describe the agenda, afterwards to outline the outcome, publication of advice to ministers, an annual press conference and an annual report),
- increasing consultation, particularly on "first of a kind" proposals,
- provision of educational material,
- research into consumer perception and food choice,

- guidance notes on commercial confidentiality,
- publication by the Food Advisory Committee of guidelines on food labelling.

The recommendations are described in more detail in the Annual Report of the ACNFP for 1990, and the changes have been in place for some years now. They proved to be satisfactory until the advent of GM soya.

6.5 The Case of the Transgenic Sheep: an Early Ethical Dilemma

In 1990 the Committee was asked whether animals from transgenic breeding programmes in which the attempted genetic modification was unsuccessful could be used as food. These sheep had been modified to carry the human gene for factor IX, a protein involved in blood clotting and needed for the treatment of haemophiliacs. The purpose of the research programme was to develop a cheaper and safer source of the factor, which had up to then been obtained from human blood, and was liable to contamination from human viruses, especially AIDS.

The gene was introduced by injection into the fertilised egg before reimplantation and rearing. Thus the introduced gene is present in all the cells of the animal, but is not active in all of them. And in this particular case, the clotting factor is released only into the animal's milk, from which it is readily purified. The process is, however, not very efficient. In many cases, the injected gene does not integrate and is degraded. In other cases, the gene is present, but not in a form in which it can work. So a large number of animals, often over a 100, are reared to obtain one animal that produces factor IX in high yield.

We were asked about the animals which contained no gene – and therefore were absolutely normal, although they had been part of an experiment – and also the animals which contained an inactive gene or only part of a gene. Could they be eaten? For they were being destroyed and this seemed wasteful if they could be eaten.

We could not think of any food safety reason why animals without any detectable foreign DNA should not be available as food. However, in considering the wider issues the Committee believed that there could be ethical and moral concerns about the food use of such animals. Would consumers object to eating an animal that had been part of a scientific experiment? And what about the animals containing an inactive human gene? Was this gene just a stretch of DNA like any other? Or was it special, because it came from a human being? Would people object to eating an animal containing a human gene? Was it even akin to cannibalism? And were newspapers going to run as a headline "Failures from genetic engineering in your supermarket" if we said yes?

Would Muslims or Jews be concerned about pork genes in lamb, and vegetarians about animal genes in plants? We did not know, but decided that it was probably a wider issue than one of pure technical safety, and suggested to the Minister that there should be wider consultation. The Minster agreed, and a small group, of which I was a member, was set up, under the chairmanship of Rev. John Polkinghorne, which consulted widely. The group received submissions from many groups, and we also talked to a number of them.

The Christians were divided. Some had no objections, but many had an uneasy concern, which they found difficult to articulate, a feeling shared by many non-Christians too, and which has been termed the "yuk" factor. The Jewish reaction was more straightforward. After all, they have been dealing with subtle issues about food for many centuries. "If it looks like a sheep, then its a sheep" was their very pertinent comment. Muslims and Hindus were much more opposed, as were the animal welfare groups and also the vegetarians.

None of the groups were moved when we pointed out that there was effectively no chance of eating the original human gene, for it was hugely diluted in the processes of genetic manipulation, and the gene inserted into the sheep was more correctly called a "copy-gene". They were even concerned if the gene was completely synthetic. They were also concerned by the "slippery slope" argument. These sheep had only one human gene in 40 000 sheep genes. But what if it was 50:50? Then I think we all would be concerned. They were also worried about labelling – they wanted consumers to have choice.

There was obviously quite widespread unease, and the Group in its Report (Ministry of Agriculture, Fisheries and Food 1993) made a series of recommendations – including the recommendation that "the first and most important requirement is for a system of labelling which permits informed choice in relation to the presence of ethically sensitive trans-genes in food". Labelling and consumer choice in 1993! If only this lesson had been remembered when GM soya was coming to the market, for by then all trans-genes had become 'ethically sensitive'. In the event, the Minister put such restrictions on food use that in practise meant that even the animals with no foreign genes would not enter the food chain.

We had learnt, in the ACNFP over those 9 years, that scientific and consumer issues should be settled simultaneously, side by side, *not* consecutively as used to be the practise. The previous approach: "First sort out the science, and then look at the consumer issues" simply does not work. A lesson we learnt first over the baker's yeast, and then over the transgenic sheep, where we realised, right at the beginning, that the question was not going to be resolved by science alone. So we asked a series of scientific questions – about chromosome fragmentation, mosaicism and the lower limits of detection of gene fragments – that we would *not* have asked but for consumer concerns.

6.6 Antibiotic Resistance Genes

The issue of a possible hazard from the spread of antibiotic resistance genes arose first in the consideration of a GM-maize variety produced by Ciba-Geigy (now Novartis). The ACNFP recommended against authorisation of this product for animal feed, its only projected use. This was because of a perceived risk associated with the transfer of an antibiotic resistance marker gene in the maize to bacteria in the gut of livestock that had been given the feed. If the antibiotic in question (ampicillin) was present in the animal feed, giving rise to selection pressure, there was a theoretical possibility of transfer of the resistance gene to humans through transfer of resistant bacteria to those in contact with the cattle. The widespread use of antibiotics in animal feed, coupled with their widespread clinical use has already led to an alarmingly high level of antibiotic resistance in bacteria that infect humans. So the debate centred on whether that figure was already so high that a very small increased risk would be of little or no significance, or whether the high level meant that no increase, however marginal, should be permitted. The ACNFP took the latter view, influenced by the potential serious outcome of an event, which although very unlikely, was not impossible (Annual Reports of the ACNFP for 1993, 1994, 1995 and 1996). The Royal Society in a statement on 'Genetically Modified Plants for Food Use' (1998) reached a similar conclusion, as did a poll conducted through the Newsletter of the International Society of Chemotherapy, where 57 % of the 198 Society members who responded opposed the use of this particular antibiotic marker gene, with a further 34 % taking the view that the risk of resistance-gene spread was low but finite (Nuffield Council on Bioethics Report 'Genetically Modified Crops: the ethical and social issues' 1999).

This recommendation was later overruled by the European Commission (EC) on a majority vote, since the maize was only to be used for animal food, and for production of starch for some processed food products, and such processing degrades the DNA so that it is no longer functional. The EC gave permission to allow marketing of the seed in January 1997, and 1000–2000 hectares were grown that year. However, in February 1998, Greenpeace applied to the French courts to overturn the issuing of the consent. The legal arguments continue.

GM tomatoes on the other hand, which contain a kanamycin-resistance gene in a form that did not cause concern to regulators, are on trial in Spain, but a commercial permit has not yet been issued.

This discussion illustrates vividly the difficulty that regulators face in making decisions that involve assessing very low levels of risk – and then in explaining them to the public. Is the risk of transfer of such antibiotic genes low, very low, or very, very low? It is ultimately a matter of judgement and all that regulators can do is to advise politicians as honestly as they can. The problem is that the public wants evidence of no risk while all regulators can ever offer is no evidence of risk.

6.7 Genetically Modified Soya

Genetically modified foods initially entered British supermarkets without any undue consumer concern. The first 2 products had been accepted – both the puree made from genetically modified tomatoes, and the cheese made by use of the enzyme chymosin made in bacteria. But the flour from genetically modified soya beans caused a huge amount of controversy, and food manufacturers have now ceased to use this product, and the other genetic modified foods that were already on the shelves in the UK were also removed, although not in the US. Why is this? Was GM soya unsafe?

The first 2 products offered the consumer both advantage and choice. Both Safeway and Sainsbury sold 170 g of the modified tomato puree at the same price as 142 g of the conventional product – because there is so much less loss in transporting the tomatoes from the field to the processing plant, and furthermore, it tasted better. Not surprisingly, the GM puree outsold the conventional product – for they were offered side by side on the shelf. The cheese made by use of the enzyme chymosin made in bacteria was marketed as suitable for vegetarians, and is currently being sold (May 2001) by the Co-op supermarket chain with the label 'Made using genetic modification and so free from animal rennet' and 'Suitable for vegetarians'.

In contrast, the flour from the herbicide-resistant soya, from Monsanto in the US, offers no obvious advantage to the consumer, only to the producer and the farmer, and the consumer has *not* been offered choice. Of course, the increased yields from this crop should stabilise or possibly even lower the price of the product, and some figures I have seen for maize suggest that roughly 25% of the increased value goes to the company, 50% to the farmer and 25% to the consumer. Even if the figures for soya were similar, the consumer cost advantage will be very small when soya makes up such a small amount of many products. So although all the evidence – including the fact that 300 million Americans have been eating it for several years without mishap – is that GM soya is as safe as conventional soya, it offers the consumer no advantage, and the scare stories might just be true. So avoiding it was a perfectly understandable reaction.

Herbicide resistant soya was genetically modified by the introduction of a gene from a soil bacterium to make the plant resistant to the herbicide glyphosate. Now we do not eat soya beans but the flour made by grinding and defatting the beans. Both the added gene and the new enzyme are degraded by this treatment, and they then will be quickly broken down in the gut. The ACNFP considered this new soya to be as safe as conventional soya, and so advised the Minister. But trust in the regulatory process had been eroded, especially by the BSE outbreak, and the claims made by Dr. Pusztai that feeding GM potatoes to rats caused pathological damage, although later showed to be untrue (see review by Gasson and Burke 2001), damaged consumer confidence so seriously that no products containing GM soya are on sale in the UK.

In summary, there are 3 main reasons why consumers have rejected GM crops and foods:

- Risks to health – slowly dying away with time as more and more Americans eat GM soya without any trace of a problem. As the OECD conference in Edinburgh a year ago (OECD 2000), said "Many consumers eat GM foods. No significant adverse effects have yet been detected on human health".
- Risks to the environment – here the jury is out and the safe conduct of the trials is very important to us all. The most recent evidence shows that GM plants do not invade the environment, but are actually poorer survivors.
- Social and ethical objections – ranging from an intrinsic objection to all genetic modification: "Going where God alone should go", to more particular concern, eating 'foreign' genes for example.

6.8 So why are Consumers so Concerned?

If GM soya is as safe as unmodified soya, and we can control adverse effects on the environment, why don't consumers want to eat it? Here is a list of reasons from the Department of Health (Bennett 1999):

Risks are generally more worrying if perceived:

- to be involuntary (e.g. exposure to pollution) rather than voluntary (e.g. smoking),
- as inequitably distributed,
- as inescapable by taking personal precautions,
- to arise from an unfamiliar or novel source,
- to cause hidden and irreversible damage,
- to pose some particular danger to future generations,
- to threaten a form of death or illness/injury arousing particular dread,
- to damage identifiable rather than anonymous victims,
- to be poorly understood by science,
- as subject to contradictory statements from responsible sources.

GM soya scores ten out of eleven! No wonder there has been trouble!

We in the UK are getting very sensitive to talk of risk, particularly as other threats to our safety recede. We live in a 'blame culture' where somebody is responsible, and culpable, for everything that goes wrong. 'Risk' isn't simple; there is a difference between those risks we choose to take, and those that are thrust upon us, and another one between risks linked to medicine and those linked to food. There is another difficulty – science can only ever say that there is no evidence of risk, while the public is now asking for evidence of no risk. That can never be supplied – with GM foods, mobile phones, or any other new technology.

So consumers want to make their own decisions, rather than trust the experts. And what are the reasons for this loss of consumer confidence? Let me suggest several:

- Scientists are no longer trusted as they once were. The BSE epidemic has of course been disastrous for confidence. Trust in the Regulatory Process has been severely eroded.
- Scientists find great difficulty in explaining, and the public in understanding, what is meant by different degrees of risk. The National Lottery – with its slogan of "It could be you" did not help either – the message is clear: even what is very unlikely may happen. So even if the risk from a new product is very low, maybe it will be me! So is the risk low, very low or very, very low?
- The public finds it difficult to know how seriously to take the points put by the many single-issue pressure groups. But the greens groups – NRPB, Greenpeace and Friends of the Earth – are probably right is saying that it is the intensive agriculture of the last 50 years that is the cause of the loss of bird life and biodiversity.
- There is widespread preference for what are seen as 'natural' processes and so the growth of organic agriculture.
- Consumers are worried about the growth of the multinational agrifood companies, and especially their effect on they developing world, and these companies have become much more aware of the issues.
- Risks are assessed differently according to the context. We will accept quite high risks when we are seriously ill, but will not tolerate much risk at all with food.

One explanation for such conflicting views is that scientists and the public work from different value systems. Scientists and technologists see novel applications of new discoveries as logical and reasonable-and characterise all opposition as unreasonable. "If only they understood what we are doing", they say, "the public would agree with us". This is not always true. Scientists and technologists are used to an uncertain world, where knowledge is always flawed, can handle risk judgements more easily, and are impatient of those who differ from them. The public's reaction is quite different, and it can be described as:

- Outrage – "how dare they do this to us?" – the way the public now regards Monsanto.
- Dread – the way we would regard a nuclear power station explosion.
- Stigma – the way the public regard food irradiation.

The net effect of this is that it is not possible to predict the way in which the public will react to a new risk by consulting just scientists and technologists, and perception of risk is now as important than any technical assessment of risk in the introduction of new technology.

So all these issues are raised by GM foods, but are they intrinsic to the technology? Rather, I believe that GM foods have become a lightning rod for many modern concerns; scepticism about the regulatory process, gusts of anxiety about our food, growing hostility to high intensity agriculture, and concern about the way in which the agrifood business has consolidated into about six companies worldwide. In all these ways people show their concerns about the way the world is going. So decisions about the future of our food are being taken in the US or in Switzerland. Consumers feel they have lost control and blame the technology, and some wish to ban it altogether.

6.9
Current Developments

There have been lay members on some advisory committees, including the ACNFP, for many years, and case studies have clearly proved their value. As the Report of the BSE Inquiry (2000) states "a lay member can play a vital role on an expert committee, and in particular can ensure that advice given by the committee addresses the concerns of, and is in a form that is intelligible to, the public". The government has stated that it agrees with this finding, but there is some uncertainty about the definition of the term "lay member". It has been stated that "It does not necessarily mean a non-scientist; indeed, some scientific background may be very useful", and "it might also refer to experts from other scientific disciplines" (House of Commons Science and Technology Committee Report on the Scientific Advisory System 2000, paragraph 69). But I remain of the view that at least one (if there are more than one) such lay person on advisory committees should be a non-scientist, for it is important that the role of a lay member is to bring an alternative perspective to the committee be maintained.

We found that the presence of a consumer or lay representative on the committee brought a number of advantages:

- their direct contribution to the debate,
- the way in which their presence has changed the way in which other members approach the issues, and
- the authentication that the consumer can honestly offer to the outside world, particularly to the consumer world.

Some have commented that the presence of a single consumer or lay representative was no more than a gesture, and that the scientists would readily overwhelm the consumer person. There is a danger that a single lay member on the committee of experts might feel isolated and inhibited from questioning the experts' view. This was not our experience on the ACNFP, as the representative was ready to say on a number of public platforms; and in practise she had an effective veto for we would never agree a recommendation against her advice. Recently however, the House of Commons Science and Technology Committee in its Report on the Scientific Advisory System (2000) has recommended (Recommendation no 44) that the "norm should be at least two lay members (depending on the size of the committee)."

Consumer representatives are now becoming the norm on Whitehall committees, but were very unusual in 1990. The presence of an ethical advisor is still unique. Yet there are still accusations of partiality. So despite that fact that the ACNFP had published its agenda, its findings and also an Annual Report – with a Press Conference – from 1990 to 2000, it was still accused of gross partiality (see Wakeford 1999 and a reply by Burke 1999). Anyone with any industrial link of any sort – a consultancy, or even funding for a research student – was judged to be hopelessly biased by opponents of genetic modification. Yet if one rules out all those who have any connections with the applications of the science, how can those left be expert enough to be of any use? The use of experts from industry was robustly defended by the Science and Technology Committee of the House of Commons (1999) as follows:

We recommend that the government rejects proposals to bar employees of biotechnology or food companies from serving on scientific advisory committees. It is vital that appointments to scientific advisory committees should continue to be made by selecting people with the most suitable and relevant expertise.

Yet the attacks continue. It is surely wiser to accept that everyone who is involved with regulation – green pressure groups included – comes from a particular background that will affect their judgement. The UK government has responded by continuing to increase the transparency and openness of the ACNFP, but leaving it otherwise largely unchanged, but it has formed a new overarching committee called the Agriculture and Environment Biotechnology Commission with 19 members from a very wide range of constituencies, from industry to green groups. The Commission met for the first time in July 2000, but it is still too soon to see how well it will work. It is certainly a commendable attempt to bring together expert technical advice from the ACNFP with a consideration of wider issues.

References

ACNFP Annual Reports (1993, 1994, 1995, 1996) Ministry of Agriculture, Fisheries and Food and Department of Health

Bennett P (1999) Understanding responses to risk: some basic findings. In: Bennett P, Calman K (eds) Risk Communication and Public Heath. Oxford University Press, Oxford, pp 3–19

Burke D (1999) If it ain't broke... Science and Public Affairs. June 1999, p 6

EU (1997a) Regulation (EC) No258/97 of the European Parliament and the Council Official Journal of the European Communities. L43: 1–7. http://europa.eu.int/eur-lex/lif/dat/1997/en_397R0258.html

EU (1997b) 97/618/EC: Commission Recommendation of 29 July 1997 concerning the scientific aspects of the information necessary to support applications for the placing on the market of novel foods and novel food ingredients and the preparation of initial assessment reports under Regulation (EC) No 258/97 of the European Parliament and of the Council. Official Journal of the European Communities L253: 1–36. http//europa.eu.int/eur-lex/en/lif/dat/1997en_397X0618.html

FAO (2000) Report of the FAO/WHO expert consultation on foods derived from biotechnology Geneva. May 29-June 2 2000. Food and Agricultural Organisation of the United Nations, Rome. http//www.fao.org/es/esn/biotech.htm

Gasson M, Burke D (2001) Scientific perspectives on regulating the safety of genetically modified foods. Nature Reviews Genetics 2: 217–222

'Genetically Modified Plants for Food Use' (1998) Statement 2/98 from the Royal Society, London

Millstone E, Brunner E, Mayer S (1999) Beyond 'substantial equivalence'. Nature 401: 525–526

Ministry of Agriculture, Fisheries and Food (1993) Report of the Committee on the Ethics of Genetic Modification and Food Use. HMSO, London

Nuffield Council on Bioethics Report (1999) 'Genetically Modified Crops: the ethical and social issues'. p 33

OECD (1993) Safety evaluation of foods derived by modern biotechnology: concepts and principles. Organisation for Economic Co-operation and Development, Paris. http://www.oecd.org/dsti/stils_t/biotech/prod/modern.html

OECD (2000) Genetically Modified Foods: widening the Debate on Health and Safety. http://www.oecd.org/subject/biotech

Report of the BSE Enquiry (2000) HC 887, Volume XI, paragraph 4.773. Also paragraph 1290. HMSO. http://www.bseinquiry.gov.uk

Science and Technology Committee of the House of Commons report (1999) Scientific Advisory System: Genetically Modified Foods

Science and Technology Committee of the House of Commons report (2000) The Scientific Advisory System

Wakeford T (1999) On peers and fellows. Science and Public Affairs, April 1999, p 29

Communication Between Scientists and Rural Communities in Zimbabwe

7

TRUST BETA

7.1 Introduction

Food is a basic commodity in our lives providing nutrients needed for human sustenance. In addition to meeting a nutritional need, a food has to be acceptable in appearance and taste. Food science is an applied science which utilizes basic sciences to study and improve the way food is handled, processed, and preserved. Research efforts in food science are expended in order to gain understanding of important biological, chemical and physical relationships that define structural and functional properties. Research is also used to develop new processing technologies, design novel ways of combining conventional ingredients, study processing effects on food ingredient performance, optimize product shelf life through reduction in growth by spoilage and pathogenic micro-organisms, and to produce healthful food products. The role of food in health and disease control is receiving great attention as evidenced by the drive towards development of functional foods and nutraceuticals.

Societies differ in the types of foods available in their environment and consumers have had to make dietary compromises in order to survive. Eating habits are influenced by food availability, nutritional needs, cultural background, quality and cost. But availability is perhaps the greatest consideration, particularly in developing countries. Consumers' understanding of food and science issues varies greatly across continents. In affluent societies of the world current scientific debates concentrate on issues involving nutritional labelling, biotechnology and genetically modified organisms in foods. In sub-Saharan Africa, the main issues revolve around food security. One approach to household food security has been to encourage production of drought-tolerant crops in drought-prone regions of semi-arid Africa. Thus, foods that feature prominently in some societies may not be significant elsewhere. Sorghum is one such cereal indigenous to Africa and grown in the semi-arid regions of the world where moisture is limited. Sorghum production is aimed at the market for human consumption in Africa, unlike in western countries, where it is used primarily for livestock feed production. The traditional methods used for processing sorghum for food have evolved over a long period of time. Sorghum meal is made into a thick porridge, a staple in sub-Saharan Africa, and sorghum-based fermented alcoholic and non-alcoholic beverages are popular in both urban and rural societies. Sorghum beer plays a significant role in African culture and is used during the performance of many ceremonies.

Zimbabwe, a country in southern Africa, is divided into five agro-ecological zones based on rainfall and soil types. Natural regions I and II, receiving >750 mm of rainfall per annum, are suitable for annual cropping and intensive agriculture. Natural regions III, IV and V, where 65% of Zimbabwe's 12 million people live, receive <750 mm of rainfall per annum. Communal areas or reserves created for the natives under the British Order in Council (1898) during the colonial era by the British South Africa Company resulted in the dual agrarian structure that divides farmers into commercial and smallholder farmers in Zimbabwe. 75% of the communal lands are located in regions IV and V characterised by low rainfall (≤500 mm per annum), periodic droughts and infertile soils. Scientists have hypothesized that the lack of food security in southern African regions that experience low and erratic rainfall, is partly caused by the use of maize, which is less suitable than more drought tolerant crops such as sorghum. A typical smallholder farmer in the communal area plants several varieties of sorghum. However, scientists in international and national sorghum improvement programmes advocate the use of a limited number of high yielding varieties for economic reasons. This raises concerns for smallholder farmers and other scientists over the diversity of germplasm available to farmers and threatens continued availability of indigenous varieties. Despite the prospects of high average yields and better food security offered by the improved varieties, farmers have continued to plant local varieties in addition to the new varieties. In general, growing modern varieties makes economic sense if ideal conditions including agricultural inputs are available throughout the growing season. However, conditions are far from ideal for many farmers and traditional varieties have evolved within their own environment and are therefore likely to be tolerant to adverse environmental conditions such as drought, pests, or diseases. To address the need for greater food security scientists in the national agricultural research systems (NARS) in the Southern African Development Community (SADC) and the International Crops Research Institute for the Semi-Arid Tropics (ICRISAT) have been working jointly on production and utilisation of drought tolerant food crops including sorghums, millets and some legumes in view of recurrent droughts in most parts of the region. Various other projects including the ecological biochemistry project funded by the McKnight Foundation (Minnesota, U.S.A.) have looked particularly at sorghum as a model food crop. Agricultural scientists, including myself, received advanced training in areas including food science, agronomy, breeding, entomology, and pathology under these programmes. As a food scientist in the NARS, my research interests have been in documenting scientific data on the food quality of sorghum and using this scientific knowledge to benefit rural societies in utilising sorghum as food.

7.2 Scientific Objectives

In the study investigating the ecological biochemistry of proanthocyanidins (condensed tannins) and related flavonoids in smallholder agricultural systems of semi-arid Zimbabwe, the scientific objectives were to determine the main sorghum varieties grown by smallholder farmers and the characteristics that farmers value in these varieties, and then characterize these varieties in terms of food quality based on physical and chemical traits. Phenolic componds in sorghum are secondary metabolites needed for defence against insects, pathogens and competitors. They are divided into the simple phenolic acids, flavonoids and polyphenols, the latter including the condensed tannins. Polyphenols, especially the condensed tannins (also known as proanthocyanidins) impart some agronomic advantages to sorghum but they also interfere with its processing and food quality. We hypothesised that the food varieties preferred by farmers (consumers) would have medium to high levels of polyphenols as such varieties have agronomic advantages. Previously, sorghum breeders have introduced new varieties only to find that they were not widely adopted by farmers because of poor consumer acceptability. The persistent use of these varieties despite aggressive promotion of new varieties could be partly accounted for by adaptive traits in the indigenous germplasm which might complement the socio-economic circumstances of smallholder farmers. Smallholder farmers, often referred to as resource-poor farmers due to inadequate resources including fertile land, animal draft power and working capital could be benefiting from the multiple uses of indigenous varieties that include food, feed and building applications. The multiple food uses of indigenous sorghums provide variety in the diet, its ability to store as seed or grain under local, cost-free technology and good agronomic properties are also advantageous.

7.3 Methodology

7.3.1 Getting Started

Planning workshops were held prior to the commencement of the project in which various stakeholder groups were represented. The project steering committee was comprised of members from the different interest groups. For example, agricultural extension officers were present as they serve to advise communities on good agricultural practices based on research findings. Some extension officers reside in the communities that our studies targeted. Members of the farmers' unions, for example, the Zimbabwe Farmers Union, represented the farmers. Some collaborating scientists had previously conducted research or had ongoing research in the same communities and were already known by villagers. Village visits were scheduled and agricultural extension officers in liaison with community leaders would organize appropriate and convenient meeting places for exchange of information. From the outset, a participatory approach was adopted in which both villagers (smallholder farmers) and scientists were jointly to implement the projects and find solutions to agricultural problems facing these communities. By involving villagers in the decision-making process, the project became theirs too.

7.3.2 Dress

In interacting with communities in rural Zimbabwe, where over 65% of the population reside, it is important that the dress code selected by the investigating scientists is "acceptable" by the villagers. Scientists easily gain trust among villagers by the manner in which they present themselves. While this might not seem pertinent to foreign scientists, the dress code is important for local scientists who want to get the most out of the rich indigenous knowledge that exists in these communities. For example, female scientists are better off clothing themselves in knee-length dresses or skirts than putting on trousers or jeans. The latter constitutes disrespect for the elders in the villages. While perceptions have been changing to tolerate western influence in cities, people in rural communities especially the elders (who are the custodians of knowledge) are very conservative. A cover cloth (dubbed *Zambia*) is a household name among Zimbabwean females and is worn in numerous gatherings, such as funerals or meetings. The cloth, a symbol of respectful dressing, serves to protect against dirt and exposure of the body during activities such as cooking. Thus the *Zambia* is important as part of the wardrobe for female scientists working with rural communities.

7.3.3 Participatory Method

In the ecological biochemistry study, a participatory rural appraisal (PRA) exercise was carried out in one of the districts which grew a large number of sorghum varieties in order to focus on farmer preferences. This was done by means of semi-structured interviews designed to collect information on varieties grown, varieties preferred and reasons for such preference. Sorghums were scored based on attributes (cooking, brewing, palatability, weevil resistance, drought tolerance, early maturing, yield, pest resistance, storage) that the farmers had identified as important. The farmers ranked the varieties from the most to least preferred. The farmers were disaggregated by gender and discussions were held simultaneously but separately for men and women. This was done to avoid the tendency of men to assume a leading role in village debates. In addition a formal household survey using a structured questionnaire was carried out seeking to establish the land area allocated by farmers to different cereals, yields and the varieties of sorghum grown. All communications were conducted in local vernacular understood by the villagers. The use of local vernacular is another way of gaining respect among villagers. However, many scientific terms are not covered in this language and these could not be easily explained. Meals and/or snacks were served during the break periods of the participatory rural appraisal exercises. In most cases, the sessions would last from mid-morning to early afternoon.

7.3.4 Laboratory Studies

After research was carried out in the field, other investigations were pursued in the laboratory. The investigations were meant to compliment farmers' knowledge and hence document the role of phenolic compounds (condensed tannins and related flavonoids) in the smallholder farming system. Sorghum varieties widely cultivated in Zimbabwe were examined for levels of phenolic compounds and kernel characteristics, to consider how to improve the acceptance and utilisation of sorghum for food in Zimbabwe and southern Africa as well as overall food security. Research was carried out to develop effective methods of processing the available polyphenol-rich sorghums. Chemical treatments were applied in milling and malting to improve processing of sorghum.

7.4 Results and Discussion

Indigenous (Consumer) Knowledge and Scientific Understanding

Smallholder farmers in Zimbabwean rural communities gave descriptions of the agronomic and food quality traits of sorghum based on many years of experience and were knowledgeable on the main sorghum varieties that they grew. Gender differences existed in the traits which farmers valued most in sorghum varieties. Men emphasised production traits (bird resistance, drought tolerance, time to maturity and seed availability), while women included traits that affected food preparation such as ease of cooking and milling (traditional pounding or mechanical milling) and quality aspects of the meal. The varieties were rated medium to high in terms of drought tolerance. Farmers (consumers) could rank the varieties in terms of preference. They were also able to differentiate three sub-groups of one sorghum variety (*Chitichi*): white, cream and brown. The brown variety was said to be bitter and mainly grown for brewing whilst the other two were used to make thick porridge – a local staple. The least preferred variety (*Chibuku*) was described as astringent and therefore not suitable for cooking. It was mainly used for brewing or marketed as grain to brewers. Farmers said that they do not feed this variety to chickens as it results in bird mortalities.

Laboratory findings complemented farmers' knowledge; scientific analyses revealed variation in kernel characteristics and polyphenol content of the sorghum varieties grown by farmers, showing that some varieties did contain proanthocyanidins (condensed tannins) which impart astringency. The latter displayed a pigmented testa containing the condensed tannins when the outer pericarp layer was removed. The surveys conducted in communal areas of Zimbabwe indicated that red sorghums without testa were being used for malting. On the other hand, sorghums containing condensed tannins were hardly used for food in rural Zimbabwe. Villagers highlighted the lack of suitable methods for removing the testa as hand pounding was inefficient. The use of a tannin-rich variety (*Chibuku* or *DC-75*) is thus strictly confined to commercial malting for opaque beer brewing as tannins are deactivated through use of formaldehyde, a treatment that is not available in village communities. Hence varieties with thick pericarp and high polyphenols were grown for marketing purposes, as simple treatments were not available for rural food processing. The thick pericarp of some varieties is desirable in traditional, manual, decortication (i.e. abrasive milling action resulting in the removal of the outer layers of the grain and therefore most of the tannins located in the pericarp and testa) but not in machine decortication. The ideal characteristics for machine decortication are a thin, white pericarp and corneous endosperm. The high polyphenol content that serves to protect sorghum against mould and pre-harvest germination affects its processing. Thus it was found that desirable agronomic properties may not be compatible with good milling quality. None of the widely cultivated Zimbabwean sorghum varieties met all the agronomic and processing criteria of high polyphenols, hard endosperm and thin pericarp.

Apart from flavour effects, condensed tannins will bind and precipitate proteins making them nutritionally unavailable. Sorghum tannins are inhibitory to the malt enzymes required for starch hydrolysis brewing. Therefore, the condensed tannins that form most of the phenolic compounds in high-tannin sorghum have to be removed or deactivated to prevent them from binding the grain proteins during food processing. In view of this, simple chemical treatments of the grain were examined in order to produce food products of low polyphenol content and therefore better nutritional and functional value so that the agronomic benefits of the tannins could be maintained.

7.5 Application of Scientific Knowledge and Difficulties Encountered

The next phase of our investigation was to apply scientific knowledge to process the available sorghums. Earlier studies have shown that some food grade chemicals have been effective in reducing the polyphenols in sorghum. We were curious about the results that would be obtained if chemical treatments were used in combination with either milling or malting. We were able to experiment using various combinations of dilute chemicals and malting or milling. Our conclusion was that the available polyphenol-rich sorghums can be processed using simple but effective treatments. The techniques reduced polyphenol levels by varying degrees. What we found most effective was a combination of dilute alkali (sodium hydroxide) treatment with malting. The polyphenol levels were reduced and high quality malt was obtained in a shorter period. We also found that the treatments were useful in roller milling but not in abrasive milling. We extended the application of these treatments to wet milling meant for starch production and concluded that the functional properties were enhanced by using dilute alkali. The question still remained about the effectiveness of alternative alkaline materials such as wood ash, that could be used at village level.

Scientific knowledge was obtained to complement indigenous knowledge and applied in addressing constraints related to processing polyphenol-rich sorghums. These findings have now been published in peer-reviewed international science journals. While this form of communication is best for disseminating results to the scientific community, it is not accessible to the smalholder farmers in rural parts of Zimbabwe.

There are a number of challenges in trying to communicate scientific knowledge to villagers and little attention has been paid to this area. The first drawback was the premature ending of the ecological biochemistry project while data was still being gathered. The funding for the second three-year phase was not secured as had been anticipated. We obtained leverage funding for continuing some of the work on sorghum food quality, however, the multidisciplinary nature of the project had been lost. There was a general lack of continuity of effort particularly in securing both human and financial resources to organise and conduct field based workshops, and develop training manuals and/or courses for the purpose of disseminating scientific findings. That was further exacerbated by the brain drain and HIV-related deaths, which have seen many research institutions losing skilled people, issues that need to be addressed within the country.

7.6 Communicating Scientific Knowledge

In the study described above, language was a major barrier to communicating scientific investigations. Villagers have terms that they have developed to describe their observations and it would be useful to develop scientific terms in local vernacular so as to enhance exchange of information. The vernacular terms could then be used to explain scientific interventions designed to address difficulties faced by consumers. Scientific knowledge could be communicated more effectively through use of terminology jointly developed by scientists and villagers. Several ways could be used to impart such knowledge to the villagers in Zimbabwe. For example, a useful dialogue was established from the beginning of the project by involving villagers in the planning stage. Surveys and participatory rural appraisals were useful in collecting data and these could also prove useful in establishing proper vernacular terms for relaying scientific knowledge. Follow-up workshops similar to the ones used during the planning stagecould be valuable for disseminating the results. "Train the Trainer" courses could be set up in which selected village members are trained in how to process sorghum in such a way as to enhance its nutritional value and eliminate anti-nutritional factors, enabling these villagers in turn to conduct training sessions in their communities using minimal resources. This approach has already proved effective with projects on soybean production and utilisation in Zimbabwe. Simple manuals or pamphlets could also be compiled, in local vernacular and used to disseminate the findings on sorghum food quality traits and processing methods.

The premature ending of projects, before scientific knowledge has been disseminated to the consumers is, in fact, an area of concern. If farmers do not receive feedback from scientists with whom they have collaborated they will become weary of communicating their agricultural problems to them. Numerous scientists have accumulated a wealth of knowledge from rural communities and their agricultural woes but have left without reporting their findings. With most projects, either the research funds do not stretch far enough or investigating scientists are more concerned with publishing their findings in international journals than giving something back to the communities that enriched their knowledge. Efforts are called for on the part of scientists to develop proposals that include effective communication of scientific findings to the rural communities who could benefit from it.

In the course of the project described here we did learn a lot about the sorghum varieties grown for food in rural Zimbabwe. The challenge still remains, however, as to how this scientific knowledge can be made known on the ground. Effective communication processes must be developed by scientists working in rural areas so that scientific results are disseminated in the less affluent societies of the world, leading to improved food security and better communities.

Reference

Beta T (1999) Processing of polyphenol-rich sorghums for food. PhD Thesis, University of Pretoria, Pretoria, South Africa

Interactive Drama and Scientific Issues

G. M. SEDDON

8.1 Introduction

This chapter describes a local project in which drama was used to promote the public understanding of science. The project was essentially a grass roots initiative that evolved out of the Norwich Meeting of the Religious Society of Friends (Quakers), where the membership includes several professional scientists and some amateur actors. In addition, there are quite a few other members who are concerned about the moral and ethical implications of recent scientific developments, as well as the lack of public awareness and discussion of these matters. This was therefore a group of people with the ideal mix of backgrounds and skills to collaborate in devising a project which used drama to achieve 2 objectives:

1. To raise public awareness of these scientific developments together with the moral and ethical implications that they raise.
2. To enable open discussion on these matters between members of the general public and professional scientists.

At the same time, the group was determined that it should not be seen to be selling a particular line on any particular issue. The underlying philosophy was to present the science objectively, and to raise the moral and ethical issues in a balanced way, with both sides of any controversy being presented fairly. In this way, individual members of the audiences could make up their own minds from a consideration of all the facts presented, and all points of view expressed.

The outcome was the development and presentation of the 4 interactive dramas summarised below. They were performed at various venues, both local and national, over the period 1996–1999.

Table 8.1 Title and content of the interactive dramas

Title	Content
A Dip in the Gene Pool	The impact of biotechnology on everyday lives
Politics and Apocalypse	The influence of man's activities on the environment
Genes Behaving Badly	The implications of the 'Selfish Gene' theory
Matters of Life and Death	Artificial methods of terminating and prolonging life

8.2 Development

The first step was to secure funding to support all the costs. The 4 productions were funded by grants from COPUS, the Cadbury Trust, and by a Royal Society/British Association Millennium Award. The budgets for each of the 4 productions varied between £ 3000 and £ 5000.

The planning process for all 4 productions was essentially the same. The group of enthusiasts met regularly over a period of 3 months to determine which particular issues were to be included in each production, and how they were to be structured and presented. All 4 productions had the same basic structure, based on 4 30-minute acts. The first 15 minutes of each act were for some form of drama that would both communicate the relevant scientific concepts, and present the ethical and moral issues. The second 15 minutes were devoted to plenary discussion among members of the audience.

Once this outline had been established, the whole project was handed over to an experienced writer to produce a script. There were then several meetings between the planning group and the writer to review the script in its various stages. The final script was handed over to an experienced theatrical director to choose a cast, and conduct the rehearsals.

Although the level of funding was very limited, it was always possible to pay either the writer or theatrical director a realistic professional fee. However, all the other personnel were unpaid volunteers, who took part in theatrical pursuits as a pastime. Fortunately, Norwich has a very strong amateur theatre company where there are always many actors and backstage technicians eager to take part in all sorts of productions in their spare time. There was therefore no difficulty in recruiting a sufficient number of good personnel.

In practice, there were 3 different writers, but they all adopted a similar approach, in breaking the drama sequences down into short review-type sketches. Each sketch made one particular point with wit and humour. Moreover, this approach was used to make both scientific and ethical points. The following excerpts from the scripts illustrate the approach in more detail.

A Dip in the Gene Pool

by *David Gwynn Harris*

This sketch is intended to raise public awareness of the scope of genetic modification. The scene is a supermarket with an actor dressed as a life-size talking carrot standing on a rostrum. A female shopper approaches, pushing a trolley.

CARROT	Carrots! Get your carrots! Lovely carrots! Loadsa vitamins! Crunchy! Delicious! Get your carrots!
SHOPPER	(With horrified reaction) What the hell are you?
CARROT	Carrot, madam. Lovely carrot. (Shopper moves closer)
CARROT	Get your carrots...
SHOPPER	Carrots don't talk.
CARROT	Not *so far*, madam. Not hitherto. But I'm an improved carrot. A genetically modified carrot. A carrot with *attitude*. I have had an implant of certain vital human genes. So now, not only am I vitamin-packed and firm textured. But I can also do my own sales promotion pitch. (Shopper looks round for help) Come on, ladies, come on! Get your lovely carrots, 40p a pound! Serving suggestion and small talk included! (Shopper passes her hand over her brow)
SHOPPER	Nonsense! I'm obviously having a nightmare. You're a chimera.
CARROT	No, I'm a carrot, madam. Very tasty carrot. No need to go calling names, now is there? (Shopper opens dictionary)
SHOPPER	Chimera, noun. A fabulous, fire-spitting monster... with a lion's head, a serpent's tail and a goat's body. Any idle or wild fancy. An organism made up of two genetically distinct tissues. *Ah!*
CARROT	Yes, well. Right you are. That's me, then. (Shopper reaches out towards carrot) Don't squeeze me till I'm yours madam. Get your carrots here! (Carrot adopts confidential tone) But I'm only one of an impressive new range of consumables. Look along the shelf there, see? Genetically-engineered lettuce. With added fish genes.

SHOPPER What? Why?

CARROT Some fish live in deep water, don't they? Deep water's cold. So they got special genes to let them survive at low temperatures. Find the gene, stick it in your lettuce, hey presto... You've got salad that stays crisper longer in the fridge. Over there, we got the tomatoes that don't go rotten, 'cos their rotting gene's been taken out. No horrible squashy tomatoes at the bottom of the pile. No waste.

SHOPPER Well, I can see the point in that...

CARROT Across the aisle on the fish counter – salmon. That's my favourite. (Even more confidentially) Got a special interest, you might say.

SHOPPER Don't tell me. It's got carrot genes in it.

CARROT Absolutely right, darling. And why? For the colour!

SHOPPER Colour! That's totally irrelevant.

CARROT Don't give me that, darling. I'm not so green as I'm... er... carrot-looking. No-one's gonna buy salmon what don't look appetising, are they?

SHOPPER But that's just cosmetic. It's outrageous.

CARROT All right, all right, try this one, then. Beetle proof potatoes. Any lady Colorado beetle who eats the leaves off of one of these gets severe constipation, and ends up with her ovaries crushed. *I'm not making this up!* No, wait, you'll love this one... Put human genes in a cow, right? What do you get? Cow's milk that's perfectly suitable for babies! Ingenious or what?

SHOPPER Yes, yes, but *human* genes? In a *cow?* No, that's appalling. I couldn't bear the...

(Enter Boffin in white coat)

BOFFIN Right, let's hold it. Stop there. This is ludicrous and irresponsible. Genes are just strings of chemicals that all living creatures have in common. Fish genes aren't bits of fish. Cows' genes aren't bits of cow. They're just bits of the same stuff, only in a different order. No animal has to be slaughtered to provide any genetic material. And quite frankly... turning an actor into something out of *The Day of the Triffids* is just pandering to the worst sort of primitive superstition and sci-fi-fantasy. It's cheap!

CARROT (Indignant) 40p a pound!

BOFFIN *Shut up!* We can enhance and conserve food stocks and increase yield by modifying the genes of our produce. This is natural. 'An art that nature makes', as Shakespeare so aptly put it.

(Enter Green)

GREEN Hey, man, where can I find the vegetarian cabbage?
(Enter Rabbi)

RABBI Oy vey, I should know? I'm still looking for a kosher beetroot!
(Enter Presenter)

PRESENTER Of course, these genetically modified products, though not all fanciful, can't yet be found in your local supermarket… Or can they?
(Produces tin of puree)
Purée from the new slow-rotting tomato strain did arrive in British supermarkets in March 1996, when Sainsbury's supermarket chain issued these guidelines for customers.
"When tomatoes ripen, they rapidly become soft and susceptible to disease and damage during harvesting. Now, Sainsbury's has worked with scientists to identify the gene that causes tomatoes to become soft when ripe. By switching off this gene, using genetic modification, the tomato remains firmer for longer, and losses during harvesting can be greatly reduced. This means savings in production, which enables Sainsbury's to sell the puree at a lower price."
In other words, responsible and informed adaptation of produce to meet a growing population's needs. Just what was envisaged by H. G. Wells' Time Traveller.
(Enter Time Traveller, reading from book)

TRAVELLER We improve out favourite plants and animals – and how few they are – gradually, by selective breeding: now a new and better peach; now a seedless grape; now a sweeter and large flower; now a more convenient breed of cattle. We improve them gradually because our ideas are vague and tentative and our knowledge is very limited. Because, too, Nature is shy and slow in our clumsy hands. But some day, all this will be better organised, and still better. The whole world will be intelligent, educated, and co-operating. Things will move faster towards the subjugation of Nature. In the end, wisely and carefully, we shall readjust the balance of animal and vegetable life to suit our human needs.
(Enter Sainsbury's spokesperson)

S PERSON While we have decided to stock the new tomato purée, we will take all future decisions on genetically modified products on a case-by-case basis. We are aware of strong ethical concerns, and have no intention of selling goods containing genes or copy genes of either human or animal origin. We understand the importance of product labelling, and think it is essential that products should be labelled to allow our customers freedom of choice.

PRESENTER Within six months, however, soya beans, rendered genetically resistant to pesticide, had found their way – untraceably – into products available in many British shops.
(Shopper looks puzzled)
So can we rely on clear labelling by suppliers to help us make an informed choice?

Politics and Apocalypse

by ***David Gwynn Harris***

Using the television programme Star Trek as a vehicle, this sketch is intended to emphasise the dangers of a market-driven economy. The scene is the control deck of a spaceship, with all the Star Trek characters on stage. The theme music from Star Trek plays over the loudspeakers.

VOICEOVER Space. The final frontier. These are the voyagers of the starship *Free Enterprise.* Their indefinite mission, to ensure continued personal survival and to avoid scuttling their own ship. Some hopes!

KIRK Full ahead, Mr Scott. Warp factor 10 overdrive.

SCOTT It's nae guid, Captain, she willnae stand it. The dilithium crystals are exhausted and we'll either shake ourselves to pieces or split our infinitives.

KIRK Well, what the hell. Let's do it anyway.

SPOCK Illogical, Captain. The power sources will burn out in precisely 9,453 days leaving us to drift helplessly with no chance of acquiring fresh supplies.

KIRK Dammit, Spock, Don't be such a killjoy. I just get a great buzz out of seeing those little lights whizzing past. Open her up, Scottie.

SCOTT Afore I do, Captain, would ye like tae buy a nice wee copper table lamp? We want to get a pool table for the engine room, so we thought we'd dismantle the air-conditioning system and sell it off as scrap.

KIRK Well, anything to improve on-board leisure facilities. We'll hardly notice much difference anyway, since the thermostat jammed with heating full on.

SPOCK Highly illogical, Captain. The consequences of reducing our capacity to regulate temperature and oxygen supply are only too obvious.

KIRK I know that, Spock, but the air-conditioning is under Mr. Scott's jurisdiction – and who am I to stop him upgrading his lifestyle to match ours on the bridge? Look! I've got a computer programme that plays intergalactic battleships with me!

UHURA Is it OK if a black woman puts a word in here?

KIRK Sure, Uhura, go ahead. This is science fiction, after all.

UHURA I have to report that the stocks of provisions in the ship's stores have run dangerously low, and have also been contaminated by leakages from the waste removal system.

KIRK Well, if they're contaminated, you better douse them with chemical pesticide to kill off any harmful bacteria. Is that clear? Seek out new life forms – and zap them.

SPOCK Totally barking mad, Captain. You may be ensuring that the foodstuffs are free of infectious bacteria, but you are incidentally rendering them potentially lethal to us as well.

KIRK Whinge, whinge, whinge, Spock. Look, wiseguy, we ditched all our provisions a while back in favour of one huge consignment of hamburgers. So, if they go rotten on us, we're in big trouble!

CHEKHOV Kyeptin!

KIRK What is it, Chekhov?

CHEKHOV I'm getting somethingk on the screen. Zere's a plynet dyead ahyead. Our sensors indicate it hez a balanced ecyology supportingk broad bio-diversity, sustainable and renewable food supplies, and low-impact technology.

KIRK What do you make of it, Bones?

McCOY Amazing. It's life, Jim… but not as we know it!

Genes Behaving Badly

by ***Justin Barnard***

The purpose of this sketch was to show the Selfish Gene Theory in operation. The scene is set on a soft summer night with the sound of flowing water, and a full moon. Two lovers enter.

VOICEOVER Don't they look romantic. Now, bring on their genes!
(Enter two genes, one male, one female. They are dressed in white denim from head to foot and wear hideous facemasks. They crouch menacingly behind the lovers.)

MAN Do you come here often?

M GENE (Whispering loudly in man's ear) Reckon she's got any assets, hanging about here? Watch out she'll have your equity in no time.

WOMAN I come here when I have the time. Listen to the sound of the water.

F GENE (Equally persistent in the woman's ear) What d'you reckon? Sound him out first. Looks alright, but he's a man! Screw you and leave you, most of them.

MAN Isn't the moon exceptionally lovely this evening?

M GENE Looks pretty fertile from here. Proposition her! Careful, though, you know what she'll be after next.

WOMAN Yes, a beautiful, constant moon.

M GENE Told you as much. Commitment. All they ever want.

MAN I would love to see you more often.

F GENE Promising. But don't forget, he's got a Y-chromosome! Pitiful.

FEMALE I would like that very much.

M GENE Steady. Could have dozens of someone else's kids already. Supporting someone else's genes – to hell with that!

MAN Do you live on your own?

M GENE Nice one. Suss her out first.

WOMAN Yes. It is a little lonely sometimes.

F GENE Good. Get his guard down first. Go for it!

MAN Sorry to hear that. If you ever want company, you only have to ask. Come for a ride in my Porsche next weekend.

F GENE That's what we wanted to hear! Porsche! Go for it now. Show him what he wants to see!

(The young woman loosens her neckline)

WOMAN Sorry about that. It's so warm to-night. I need some air!

M GENE Did you see that! Lucky she doesn't know he traded in the Porsche for a Skoda last week.

MAN Oh don't stop there. I feel the same way. It's very hot.

(He begins to unbutton his shirt)

F GENE Stop him. He's taking advantage!

WOMAN Hold on a minute before you do that.

(She produces a diary)

Do you mind if we make a few dates in advance?

M GENE I told you! Commitment! He'll never go for it. Fertilise her and be off!

MAN (Rounding on his gene) Shut up you! What if I *want* to be faithful?

M GENE You speaking to me?

MAN	Yes I am. Don't you realise? You may have a better chance of being passed on to the next generation if I stick around and care for her.
M GENE	That's novel. Never thought of it like that. OK. We'll have a bit of fun in the meantime!
MAN	Fine. I'm yours, for as long as you wish, my darling.
WOMAN	That moon is so beautiful!
MAN	It is my love!
F GENE	(To the woman) Don't trust him an inch. He's a man! (The woman knees the man in the groin. Black out)

Matters of Life and Death

by ***John Mangan***

The purpose of this sketch is to illustrate the hazards that face each individual sperm on its way to the uterus. The sketch 'Sperm Wars' is a take-off of the film 'Star Wars' and, to the theme from Star Wars playing over the loudspeaker, the scene opens with a number of rebel warriors piloting two fighters in formation, (in other words, sitting on stools pretending to fly).

LUKE	Red Leader from Luke Spermswimmer. Prepare for launch. (There is a very rude noise and the pilot sperm react as though catapulted)
R LEADER	This is Red Leader. Keep tight formation until we exit the launch tube.
BLUE	Red Leader. I've been ejaculated backwards. Can't make it. Breaking off...
LUKE	She's been ejaculated into the bladder. She's... (An explosion is heard off stage) Blue!
R LEADER	Concentrate Red Flight. She knew the risks.
SHORTY	Don't worry, Luke. There's over 20 million of us. One's bound to make it.
R LEADER	Entering cervix now! (The pilots bend sideways then straighten)
RUSS	Red Leader. My guidance system has gone haywire. I'm going in the wrong direction.
R LEADER	Russ, I've got visual on you... You're...you're...
RUSS	What is it? For God's sake, tell me!

R LEADER You're malformed, Russ. Your tail is half the length it should be. You have all the wrong chromosomes.

RUSS You mean I'm not viable, Red Leader.

R LEADER I'm sorry...

RUSS Don't be. I can take it. Don't worry guys. I'll do the right thing.

(He presses a button and vanishes. An explosion is heard)

R LEADER What a great guy ... he might have been!

CHUCK Red Leader. I'm picking up a muscle spasm in the cervix. Brace yourselves.

(The pilots rock back and forth)

CHUCK Losing power. I can't... make it. Good luck, Red Flight.

(Chuck vanishes. An explosion is heard)

SHORTY Chuck!!

R LEADER We just lost half the flight. Cervical mucus ahead.

(The flight starts moving in slow motion)

LUKE Wow. This is really thick. Shorty? Shorty?

SHORTY I'm... stuck... Luke. The whole wing is immobile. You'll have to do it for all of us, Luke.

(Shorty vanishes. A number of close order explosions are heard)

R LEADER Through the cervical mucus... Now! I'm picking up hostile antibodies. Watch your back, Luke.

(Sounds of X-Wing Fighter Scream. Darth Vader appears)

D VADER The Force is strong in this one. Antibodies. Use chemical weapons.

(Ray gun noises)

R LEADER They're increasing the acidity levels. I'm breaking up...

(Red Leader vanishes)

D VADER Give up, Luke. You're the only one left. This ovum is mine.

LUKE Who are you?

D VADER I am your father's sperm.

LUKE WHAT?

D VADER Sperm can survive three days in the cervix, Luke.

LUKE My own father? That's disgusting!

D VADER Mother Nature doesn't have ethics, Luke. A viable sperm is all she needs. Look!

(A huge egg appears)

LUKE It's huge!

D VADER One egg per month viable for only a few days per month. See why you must fail in your puny quest!

LUKE I'll never do it. It's a thousand times bigger than I am.

D VADER	I created your father, Luke. My track record is clear. Now, I will create your son!
LUKE	No. Only one of us can get inside and I'm younger, fresher…
D VADER	Unproven…
LUKE	He's right. What if the genetic package I carry turns out to be firing blanks? (Voice of Alec Guinness)
A G VOICE	Follow your instincts, Luke. Every sperm knows what to do. Let go of your pre-conceptions. Yes. Let go of your pre-conception pre-conceptions.
LUKE	Diving … now! (Sound of two fighters diving and Darth Vader's lasers firing)
A G VOICE	Follow your instincts, Luke. (Darth Vader reaches the bubble first. He can't get in)
D VADER	Curses. My genetic packet is too old! (Luke penetrates the bubble. Darth Vader bounces off into the darkness)
D VADER	It isn't over yet, Luke. A lot can go wrong in nine months. (Blackout)

8.3 Presentation

Although it was possible to hire costumes, good quality sound equipment, and rudimentary stage lighting, the budgets for each production were not sufficient to cover the construction of elaborate scenery. The stage sets were essentially bare with just a few props such as tables and chairs.

All the sketches had been deliberately written with these limitations in mind, and all the artistic directors and casts were well used to working under minimalist stage conditions. In the event, the lack of scenery did not detract from the performances in the slightest. The focus of attention was always on the actors and what they were doing and saying. Scenery would have been unnecessary, and possibly distracting.

The absence of scenery also allowed performances to take place in venues which were not primarily designed as theatres. For example, there were performances in a lecture theatre with a dais somewhat lower than a theatrical stage. There were also performances 'in the round'.

The productions made much use of humour, and often drew on popular culture. Another vehicle for enhancing lay understanding of scientific knowledge was the use of analogy in visual aids. For example, the explication of genetic code was compared with communication of a Shakespeare text in Morse code by a flashing torch. The double helix structure of DNA was demonstrated as a circus act involving two towers of acrobats sitting on each other's shoulders. The acrobats wore different coloured gloves to represent different bases, and acrobats in the different towers held hands to represent the pairing of the bases in parallel chains of DNA. These demonstrations made the productions entertaining as well as informative, and marked them out as a distinctive means of enhancing understanding and stimulating public debate on serious and pressing issues arising out of advances in science and its implications.

8.4 Discussion Sessions

The discussions were always lively, with no shortage of participants, though sometimes a warming up period was required. Secondary school students from the age of 14 up were particularly vocal, uninhibited by the occasion, and the standard of their contribution was very high. Although discussion could often have extended beyond the 15 minutes allocated, it was not allowed to do so as the chairman always stuck strictly to the schedule for the evening. A variety of views were expressed but actual conflict never developed. *A Dip in the Gene Pool* elicited much concern amongst audiences regarding the medical aspects of applying genetic knowledge, such as the making available of individuals' genetic profiles to insurance companies, and confidentiality. There was also vociferous opposition among some to the genetic modification of plants and food crops because of uncertainty about long term effects. Again, discussions during *Genes Behaving Badly,* showed that audiences were keen to discuss issues of practical application and how these might affect human lives, rather than the concept of the "selfish gene" itself.

The discussion sessions were conducted by a chairman. Although the variation in seating layouts in different venues did not seem to affect the quality of discussion, those occasions conducted in the round worked particularly well as this arrangement meant that people could see one other while discussions were going on. For 3 productions the chairman was not a member of the cast, and most, but not all, of these chairmen were scientists. However, it was found that the chairperson's background was less important than his or her ability to facilitate discussion and help the event to move forward; it was important to be perceived to be in charge.

The fourth production experimented with using a chairman who was also a member of the cast. In this case, the whole production was presented as a television chat-show, in which the character Randy Curlew was a cross between Jerry Springer and Rikki Lake. This latter format was the least successful because many members of the audience could not dissociate the person who was chairing the discussion from the character who was taking part in the sketches. As a result, they were inhibited from making serious points, and tended to make flippant replies to the questions from the preposterous personality who was thrusting the microphone into their faces.

Politics or Apocalypse

This discussion took place after scene in a court room where 'Everyman', a man in the street, is put on trial before God to answer charges of complicity in

1. *Depletion of global resources*
2. *Poisoning the atmosphere, land and sea*
3. *Destruction of the environment*
4. *Extinction of species*
5. *Erosion of habitat*
6. *Exhaustion of agricultural land*
7. *Unfair exploitation of the world's labour potential*

Evidence for the prosecution and defence was presented in the sketches. In the discussion session, the audience took on the role of the jury. What follows is an edited version of one audience's deliberations.

WOMAN 1 He is guilty because he could be recycling his waste to a large extent, picking up tin cans in the street, or whatever. He could also be buying *Traidcraft* or *Tearfund* food to live on.

WOMAN 2 He is guilty. He could use many items that he throws away. He could reclaim items which are no longer used for one purpose to become raw materials for another purpose. He could reduce his consumption and make things last longer.

CHAIRMAN If he is guilty on those grounds, how many of us actually manage to do all of that. If not many of us manage to do all that, can you really hold the individual responsible? Surely, that's down to education, the government, the way that big business is advertised, and so on. I don't think it's fair to hold one person responsible.

WOMAN 3 We are each responsible for our own actions. Therefore Everyman is responsible for his own personal guilt. We are all more or less guilty of the charges brought against him tonight.

WOMAN 4 I cannot really say he is absolutely guilty. Although we are all guilty of all the things he has been accused of, there are other people or other things that are more guilty. Materials are used to make profits. Profits are used to give many people high standards of living, but not the people who produce the raw materials. This does not apply only to the Third World. There are many people in Europe that are not really enjoying the fruits of this wonderful society. I think education has a lot to do with the guilt that is placed upon most of us.

MAN 1 If Everyman recycles his packaging for instance, industry will have to sell more packaging, because industry has to expand. So, they will have an advertising campaign to sell more packaging. If they can't do that, they will have to use the capital in some other way. If everybody uses fewer materials, they will use capital some way that does not depend on individuals, such as putting it into the arms trade. So, whatever Everyman does, we cannot win.

WOMAN 5 I think we are being very simplistic about all this. I think we don't all realise what pressures we are under. I think the power of advertising does not give us time to question what is important. I think we are being very smug if we do these things.

WOMAN 6 I think we need a lot more encouragement. We have mentioned education already, but we also need encouragement in terms of what is provided for us to recycle – the provision of bins and so on in different colours.

MAN 2 I think it is wrong to frame this in terms of guilt. If you want Everyman to change, you must show him the benefits of simple living – and the joys.

WOMAN 7 Everyman doesn't really take part in changing his perceptions, or changing his impact on the world, because he is frightened of being different, and he is frightened of being wrong. I don't think you can condemn him for being frightened of being different or wrong.

WOMAN 8 You cannot really find someone guilty, unless they intended to commit a crime.

WOMAN 9 While these are all valid points about what we consume, I thought that one of the wider issues being considered was First World against Third World starvation and poverty, the real starvation and famine that we see in the Third World. I think that we are all getting a bit too narrowly concerned about how many plastic bags we use, and not looking at the fact that everyone in this society is so much more wealthy than the sort of people we see in Rwanda.

CHAIRMAN What could Everyman have done to address that issue?

WOMAN 9 I suppose make our political leaders aware of the fact that we care about what happens in the Third World, and that we think it is important.

WOMAN 10 Like most of the people here, I am aware of how complex all this is. In a way it is not an excuse to say, 'Well I didn't realise what I was doing'. We should know what we are doing!

WOMAN 11 I don't think anyone of us is guilty because we don't feel in control, or are in control, of the machine that is taking us forward. In the end, individuals can only do small acts.

CHAIRMAN Can I ask you who you think is at the controls?

WOMAN 13 I don't think anyone is in control. I think greed has got us into a snowball. There is no way to get out of that.

MAN 3 If we are to change behaviour, it must not be out of fear of punishment. I certainly think we should be aware of the consequences of our actions, but it is far more important to change out of a wish to be positive, and to do the right thing.

WOMAN 14 I think a vast number of people in the world have good will and want things to change, but we will have to put political pressure on the people who govern us before anything will happen. I don't think it is enough just to re-cycle bits of plastic, because they will just make more bits of plastic for us to re-cycle.

MAN 4 We must be careful to distinguish what's a humanist issue and what's an environmental issue. I know the two are completely intertwined but we have to ask ourselves what we are putting first. Should we put the humanist issue first, or the stability of the global climate first.

CHAIRMAN So, we should be thinking about what impact we have in each issue?

MAN 5 Yes. One way is to measure the consequences of our actions against their effect on the climate. Traditional ethics tend to measure the effects of our actions on people, but rarely do we consider other members of the natural world.

WOMAN 15 I would like to say that it is God's fault. In terms of the Judaeo-Christian religion, the human is seen as akin to God. Animals and the world are just for our pleasure, and that's what we have made them.

WOMAN 16 Part of the problem is that the qualities we needed to survive when we were coming out of the trees, are the qualities that we are left with. They are not the qualities that are suitable for today's life, but I don't know how we can get rid of the aggression, that is built into us.

WOMAN 17 Why can we not devise a system that limits the amount of property we own. I am limited by law in the amount of noise I can make, but I can own half the county. That seems crazy.

CHAIRMAN Let us now take a vote as to whether Everyman is guilty or not.

(Three people voted 'Guilty'; forty voted 'Not Guilty')

8.5 Audiences

The nature of the publicity campaign varied from venue to venue. The ticket price was kept deliberately low, i. e. £ 3, in order not to deter anyone from attending. In Norwich, attractive posters were distributed through the publicity network of the Maddermarket Theatre and the University of East Anglia. Mail shots went out to all local schools and religious organisations. Interviews and announcements were arranged on local radio stations. When performances took place further afield, e. g. in Cambridge and Edinburgh, the publicity was entirely in the hands of the host organisation.
The effectiveness of this publicity varied greatly. There were sell-out audiences of 300 for some of the performances in Norwich. Indeed, for each production in Norwich, there were always 2 runs, one in March as part of National Science Week, and one in October. In contrast, there were only twenty people or so in the audiences for each of the performances at the Edinburgh International Science Festival!
The audiences always contained a good number of students. In the main, the other members of the audiences were educated and middle class. Non-scientists usually outnumbered scientists but, as might be expected, it proved difficult to attract people who were not already aware of and concerned about the issues to some extent, other perhaps than any students who came in a party with their school and had not yet begun to think about such questions.
One effective means of spreading wider the kind of opportunity offered by such productions would be to video them – though video can never capture or create the immediacy of live theatre. Discussions could be omitted so that video audiences could generate their own. Such videos could be distributed to schools through an agency such as the Association for Science Education, though creating stocks of videos, marketing and distribution would incur considerable costs, and accessing other, non-school, groups would be difficult.

8.6 Overview

As no formal evaluation of these productions was carried out, it is not possible to give an objective appraisal of the whole project. However, it is a well-known 'showbiz' maxim that you are only as successful as your last production. So far as the successive productions in Norwich are concerned, the project has been able to sustain a faithful following of schools, who have brought their students to each production. Moreover, there has been a great deal of complimentary feedback from many different directions. There is no doubt

that the quality of productions has been of a good standard, and has created a very favourable impression with all the audiences. From that point of view, everyone involved in the project feels that the whole project has been very worthwhile.
The most difficult task of all has been marketing the productions and attracting large numbers of people from the whole spectrum of society. On such a low budget, the publicity cannot have a really significant impact on all areas of the general public. The important task in future ventures of this kind should be in devising methods of communicating with a much bigger and wider section of the general public.

8.7 Other Similar Enterprises

The interactive dramas described here were among the first of several such initiatives in different parts of UK. It therefore seems that this medium of communication has indeed been found to be a well-received and effective vehicle for raising awareness and increasing understanding of science and ethics.
A play called *Footprints*, commissioned by the British Association and financed by NESTA (National Endowment for Science, Technology and the Arts) with Science Year funding, is currently being devised by the Amoeba Theatre Company. It will explore genes and evolution, how they have affected the past, influenced the present, and have implications for the future. During the course of the play, the audience will be asked to make decisions that will directly alter the plot. The production will tour the country for a target audience of 12 – 14 year-olds, their parents and teachers.
Another theatre company, Y Touring, has considerable experience of creating several productions which incorporate audience discussion. In their case, the members of the cast have stayed in character to field questions during the discussion which has followed each play. The Wellcome Trust's Medicine in Society Programme, and the Office of Science and Technology's public understanding of science initiative have commisioned Y Touring to tour with four *Theatre of Debate* productions. Their themes have been xenotransplantation, genetic selection, cloning, and the biological basis of mental illness. Each play was written in consultation with scientists, doctors and patients, and is supported by a resource pack.
Audiences have included young people, teachers, school governors, parents, scientists and other members of the general public. Productions have been well received by both the scientific and medical press and by broadsheet newspapers. An independent evaluation of *The Gift*, the play about genetic selection, came to a very gratifying conclusion:

The participating teachers have found that this project has had a very positive impact on the students. The students are seen to have gained more knowledge on genetics than they would have done in more formal, straightforward lessons.

8.8 Conclusion

Interactive drama can break down the barriers between art and science, as well as between scientists and lay public. It is an effective way of increasing understanding of certain areas of science and awareness of the social and ethical issues to which they give rise. The productions do not have to be elaborate to be effective, but they do require an appropriate level of funding in order to widen and maximise the potential audience. It is important, however, that such dramas are soundly informed, and offer multiple views.

Acknowledgements

The author would like to thank COPUS, The Cadbury Trust and the Millennium Commission for their grants. He would also like to thank David Gwynn Harris, Justin Barnard, and John Mangan for permission to use extracts from their scripts.

Dr. Rainer Wild-Stiftung – Holistic Approach to a Healthy Diet

The Dr. Rainer Wild-Stiftung was founded in 1991 by Dr Rainer Wild, an internationally-practising businessman and honorary professor at the University Stuttgart-Hohenheim, Germany.
This Heidelberg-based foundation made it its objective to promote healthy nutrition in the individual. To this end, the Foundation carries out its own research, promotes relevant scientific work that serves its goals, and provides further education and training specifically for mediators employed in the field of nutrition.
Since a healthy diet is more than counting calories and includes more than the examination and composition of food and food products, a modern holistic approach to a healthy diet is foregrounded. In such an approach, the individual and their individual wishes and needs are focused on. Not only what, but also how, why and where we eat plays an important role.
Consequently, the Dr. Rainer Wild-Stiftung not only concerns itself with purely scientific issues, but also accounts for the cultural and social aspects of nutrition in its work. In all its activities, the Foundation for Healthy Nutrition always seeks to promote interdisciplinary dialogue.
The work of the Dr. Rainer Wild-Stiftung includes the following activities:

1. The "Heidelberg Nutrition Forum" (*Heidelberger Ernährungsforum*) is a series of conferences hosted by the Dr. Rainer Wild-Stiftung. At regular intervals, the forum offers the possibility for representatives of the sciences, the industry, and other areas to discuss issues of a healthy diet in an interdisciplinary manner.
2. Since the end of 1998, the Springer-Verlag has been publishing the series "Healthy Nutrition" (*Gesunde Ernährung*). It is one of the few scientific series which consciously relies on a broad readership interested in a healthy diet.
3. Aside from its own research, the Dr. Rainer Wild-Stiftung initiates projects dealing with the issue of a healthy diet, in part, in close co-operation with selected partners.
4. The International Association for Research on the Culture of Eating is an interdisciplinary circle of experts in natural and cultural science. Its offices

are with the Dr. Rainer Wild-Stiftung. Symposia on "Eating Culture" (*Esskultur*) are regularly held and the resulting reports are published.

5. With the Dr Rainer Wild Award, every two years the Foundation honors outstanding achievements in the area of nutrition in accordance to the philosophy of the foundation. Corresponding to the importance of the issue, the award includes a prize of € 15,000.

This range of activities highlights the fact that the Dr. Rainer Wild-Stiftung is intensively concerned with the complex topic of "nutrition". With its work, the Foundation strives to enable the individual to take responsibility for their own health and to consciously maintain and promote it. The way we eat cannot be separated from the way we live.

For further information, please contact:

Dr. Rainer Wild-Stiftung
In der Aue 4
69118 Heidelberg
Germany
Telephone: +49 (6221) 89 98 0
Fax: +49 (6221) 89 98 40
e-mail: info@healthy-nutrition.org
http://www.gesunde-ernaehrung.org

About The Authors

Peter S. Belton
School of Chemical Sciences, University of East Anglia,
Norwich NR4 7TJ, UK, p.belton@uea.ac.uk

Professor Peter S. Belton is currently Professor of Biomaterials Science in the School of Chemical Sciences at the University of East Anglia. Before that he held various senior management posts at the Institute of Food Research in the UK. He is also a Vice President of the UK Institute of Food Science and Technology. His current technical research interests include structure function relationships in seed storage proteins and the applications of spectroscopy to biomaterials.

Teresa Belton
School of Education and Professional Development, University of East Anglia,
Norwich NR4 7TJ, UK, t.belton@uea.ac.uk

Dr. Teresa Belton has a first degree in art history. Currently she combines academic research with non-academic work as Learning Officer for a rural regeneration partnership. Her research experience is in qualitative studies in education and health, and she is currently Visiting Fellow in the School of Education and Professional Development at the University of East Anglia.

Trust Beta
Department of Food Science, University of Manitoba, Winnipeg, Manitoba
R3T 2N2, Canada, betat@ms.umanitoba.ca

Dr. Trust Beta has been involved in agricultural research and training aimed at improving household food security in rural communities of Zimbabwe through her activities in the national agricultural research systems (Department of Research & Specialist Services and University of Zimbabwe) in collaboration with international institutions. She is currently working at the University of Manitoba in the Department of Food Science.

Derek Burke
13 Pretoria Road, Cambridge CB4 1HD, UK, dcb27@herems.cam.ac.uk

Professor Derek Burke was Vice-Chancellor of the University of East Anglia from 1987 to 1995. Before that he was Vice-President and Scientific Director of Allelix, Toronto, Canada, and Professor of Biological Sciences in the University of Warwick. He was chairman of the Advisory Committee on Novel Foods and Processes from 1988 to 1997, a member of the Committee on the Ethics of Genetic Modification and Food Use, a member of the Nuffield Council on Bioethics' Working Party on Genetically Modified Crops, and a member of the Science, Medical & Technology Committee of the Church of England's Board for Social Responsibility, and chairman of the Working Party that produced "Cybernauts Awake! Ethical and Spiritual Implications of Computers, Information Technology and the Internet". He was a Specialist Advisor to the House of Commons Select Committee on Science and Technology, and also of the EU-US Consultative Forum on Biotechnology.

Lynn Frewer
Institute of Food Research, Norwich Research Park, Colney,
Norfolk NR4 7UA, UK, lynn.frewer@bbsrc.ac.uk

Dr. Lynn Frewer is currently Head of the Consumer Science Group at the Institute of Food Research in Norwich. Her current research interests include the psychology of risk perceptions and attitudes, the influence of the media on risk perception, public reactions to genetic modification and other emerging technologies, the impact of trust on the effectiveness of risk communication, developing methodologies for fostering public participation in strategic development of food technologies, and health psychology. She has published over 100 publications, mostly focusing on issues associated with risk perception, risk communication, and public involvement in strategic decisions regarding risk management. She has been involved in numerous international and national consultation groups concerned with risk communication.

Anne Murcott
South Bank University, 103 Borough Road, London SE11 OAA, UK,
murcota@sbu.ac.uk

Professor Anne Murcott holds an MA in Social Anthropology from the University of Edinburgh and a PhD in Sociology from the University of Wales. Author/editor of books and numerous articles in sociology on various aspects of food and eating, diet and culture, and also health, she served as Director of the Economic & Social Research Council (UK) Research Programme "'The

Nation's Diet": the social science of food choice' (1992–1998). Her most recent work includes a co-edited volume entitled *Developments in Sociology* published by Pearson in 2001. She is now honorary professor at City University, London, and at the Institute for the Study of Genetics, Biorisks & Society (IGBiS), University of Nottingham, and is Professor Emerita (Sociology) South Bank University, London.

Jacquie Reilly
Mass Media Unit, Department of Sociology, University of Glasgow,
Adam Smith Building, Glasgow G12 8RT, UK, j.reilly@socsci.gla.ac.uk

Dr. Jacquie Reilly is a Research Fellow at the Mass Media Research Unit, University of Glasgow. She has over ten years experience of research into the production, content and reception of risk information in relation to food safety issues. Publications include: Miller, D and Reilly, J (1995) 'Making an Issue of Food Safety: The media, pressure groups and the public sphere' in Donna Maurer and Jeffrey Sobal (eds.) Food Eating and Nutrition as Social Problems: Constructivist Perspectives, New York: Aldine De Gruyter. ISBN: 0 202 30507 4. Reilly, J and Miller, D (1997) Scaremonger or Scapegoat? The Role of the Media in the Emergence of Food as a Social Issue, in Caplan, Pat (ed.) Food, Identity and Health, London: Routledge. Reilly, J (1998) 'Just another Food Scare? Changes in public understandings of BSE', in Philo, P. (ed.) Message Received, London: Longman. Reilly, J (1999) 'The Salmonella-in-eggs crisis in Britain', in Applebaum, M (ed.) Alimentation, Peurs et Risques: Ouvrage sous la Direction, Paris: Observatoire Cidil de L'harmonie Alimentaire.

G. M. Seddon
Centre for the Advancement of Science and Technology Education,
31 Westwick St, Norwich NR2 4TT, UK, g.m.seddon@uea.ac.uk

Dr. Malcolm Seddon is qualified both as a chemist and an educational psychologist. He combines both disciplines in carrying out research and development projects in science education at all levels. He teaches chemistry at undergraduate level and the methodology of educational research to experienced science teachers on advanced courses. He has held academic posts at the University of Birmingham and the University of East Anglia. He is the founding director of the Centre for the Advancement of Science and Technology Education.

Subject Index

J

K

L

M

N

O

P

V

W

Z

Printing: Mercedes-Druck, Berlin
Binding: Stein + Lehmann, Berlin

Book series edited by the Dr. Rainer Wild-Stiftung

Bergmann K., München, Germany

Dealing with Consumer Uncertainty

Public Relations in the Food Sector

The current practice of communication in the nutritional economy often produces significant uncertainty in a large fraction of the population. Efficient and comprehensive publicity by entrepreneurs on the industrial production of foodstuffs needs a new concept for communication between producers, processors, wholesalers, retailers, and end users. Without overgeneralizing, the author explains what makes the consumers uncertain and which consequences this uncertainty has for their nutritional behavior. The main aim of this book is the empirical explanation of the connection between the uncertainty concerning the health value of industrially produced foodstuffs and the behavior of consumers in relation to information. It shows how consumers currently perceive the publicity activities of the food industry and what their needs are as far as information is concerned. The practical consequences derived from the empirical results are comprehensibly described.

2002. XX, 218 pp. 15 charts, 64 tabs., 20 graphs, Hardcover
3-540-42529-2
EUR 49,95
Recommended Retail Price

Book series edited by the Dr. Rainer Wild-Stiftung

Grimme, L.H., University of Bremen, Germany
Dumontet, S., Ordine Nazionale dei Biologi, Rome, Italy
(Eds.)

Food Quality, Nutrition and Health

5th Heidelberg Nutrition Forum/Proceedings of the ECBA – Symposium and Workshop, February 27 – March 1, 1998 in Heidelberg, Germany

Responding to the dramatic scientific and technological developments in the agro-food sector and to the enormous public concern about novel food production and novel food ingredients this volume focusses on defining, classifying and reassessing the quality of food towards human nutritional needs aimed at health. It is designed for all those actively involved in the food sector and for interested lay persons and responsible consumers interested in getting information about the driving forces of the present and future food market, the food industry, and the food policy and the consumer association.

2000. XI, 214 pp., Hardcover
3-540-65997-8
EUR 39,95
Recommended Retail Price

CPSIA information can be obtained at www.ICGtesting.com
Printed in the USA
LVOW100207231012

304022LV00006B/25/A

9 783540 437437